SDGs and Textiles

Series Editor

Hafeezullah Memon, International Institute of Silk, Zhejiang Sci-Tech University, Hangzhou, China

The book series "SDGs and Textiles" addresses the strategies to achieve sustainable development goals (SDGs) in the present, past, and future. It presents books about the present and future policies of textile ministries of different countries, and books related to sustainability education around different parts of the world in the textile sector. Moreover, it would welcome the conference proceeding related to SDGs and Textiles. The series would cover books comparing the sustainability and SDGs of different institutions and countries. The individual book volumes in the series are thematic. The goal of each book is to give readers a comprehensive overview of a different area of sustainability in the textile sector. As a collection, the series provides valuable resources to a broad audience in academia, the research community, industry, and anyone looking to expand their knowledge of SDGs and Textiles.

Textiles and life are together – life cannot be separated from textiles as it is the most important need for human beings after food. In 2015, the United Nations General Assembly proposed 17 interlinked global goals to be achieved by 2030. Since then, academia and industry have paid much attention to achieving these goals. Textile found its close relation with almost all of these 17 goals.

SDG 1 - No Poverty: Poverty would never be overcome by a charity only; it is essential to develop people's skills to have a better and wealthy life. Thus, the textile can be considered an excellent discipline to achieve this goal by creating jobs and small and medium businesses.

SDG 2 - Zero Hunger: Through the effective utilization of advanced application of Agrotech Textiles, it is possible to have higher crop yields and save crops from rough weather, unexpected rains, floods, insects, etc.; thus, geotextiles play an essential in achieving this goal of sustainable development.

SDG 3 - Good Health & well-being: There has been much health consciousness after Covid19, and medical textiles assist in getting good health and well-being.

SDG 4 - Learning & Education: Textile or fashion has remained a significant discipline for societies for ages, and there has always remained much to explore in this field. Textile-related universities may play a vital role by offering free access to their education resources, training and spreading information among the locals.

SDG 5 - Gender Equality: The textile sector is one of the industrial sectors that accepted gender equality long ago; in particular, the garment sector has more females than males. Thus, the textile sector has been doing gender equality. Moreover, there has been a recent trend for Gender Neutral Clothing, which need worth studying and may further assist gender equality.

SDG 6 - Clean Water & Sanitation: Textiles could be achieved through filtration, and of course, textile is one of the critical materials for filtration.

SDG 7 - Affordable & Clean Energy: With the recent advancement in material science and engineering, the textile sector has come on the front for, not only by using this clean energy during textile production but also by assisting the production of this clean energy, either in the form of wind turbines blades made of textile composites or by energy harvesting from T-Shirts, etc.

SDG 8 - Decent Work: Recently, there has been much attention that the textile workers are not paid well, labor rights are not cared about, etc.

SDG 9 - Industry and innovation: Textile Industry always follows innovation; the textile companies that do not chase innovation cannot survive in the market.

SDG 10 - Reduced Inequalities: Getting better life and well-being would help reduce inequalities in the textile industry.

SDG 11 - Sustainable Cities: Sustainable Textile Cities through Buildtech and transport textiles.

SDG 12 - Consumption and Production: Textile and garment consumption and production all come under.

SDG 13 - Climate Action: Oekotech or Ecotech Textile, waste management of textiles are upfront to achieve this goal of sustainable development.

SDG 14 - Life Below Water: Mitigating microfiber waste in rivers and oceans may come under the context of it. There has been much attention on this subject after passing the bill at the parliament level of the UK.

SDG 15 - Life on Land: Geotech or Geotextiles studies life on land.

SDG 16 - Peace, Justice, and Strong Institutions: Protective textiles are doing their best to achieve peace, justice, and strong institutions.

SDG 17 - Partnerships for the Goals: The application of textiles to achieve sustainable development goals is only an example. In all textiles sectors, combined efforts of all the goals are essential to achieve true sustainability.

Muhammad Qamar Khan · Yasir Nawab ·
Ick Soo Kim
Editors

Garment Sizing and Pattern Making

Springer

Editors
Muhammad Qamar Khan
Department of Clothing
National Textile University
Faisalabad, Pakistan

Yasir Nawab
School of Engineering and Technology
National Textile University
Faisalabad, Pakistan

Ick Soo Kim
Faculty of Biomedical and Textile Science
Shinshu University, Ueda Campus
Ueda, Nagano, Japan

ISSN 2948-1236 ISSN 2948-1244 (electronic)
SDGs and Textiles
ISBN 978-981-97-7682-5 ISBN 978-981-97-7683-2 (eBook)
https://doi.org/10.1007/978-981-97-7683-2

This Springer imprint is published by the registered company Springer Nature Singapore Pte Ltd.
The registered company address is: 152 Beach Road, #21-01/04 Gateway East, Singapore 189721, Singapore

Introduction

This book investigates the best possibilities of garment constructions through proper pattern techniques which can enhance the efficiency of organization and minimize the wastage of materials within sustainable developments. It also enhances the importance of sizing system for brands and suppliers for development of garment fit, garment comfort and durable garments. This book will also empower the DIT Designers for the best entrepreneur. It also covers the many practical areas as like mentioned below:

Education and Skill Development: This book can serve as a valuable educational resource for students, aspiring designers, and professionals in the fashion and garments industry. It can provide detailed instructions and insights into the principles and techniques of creating garments that fit well and look aesthetically pleasing.

Improved Garment Quality: Understanding sizing and pattern making is crucial for creating garments that fit properly. This book can help designers and pattern makers develop the skills needed to create high-quality garments that flatter the wearer's body shape.

Standardization and Consistency: Many industries require standardized sizing to ensure consistency in clothing production. This book covers sizing and pattern making can help establish and maintain such standards, which is crucial for mass production and retail.

Innovation and Customization: The knowledge gained from a book on garment sizing and pattern making can empower designers to create innovative designs and customize clothing for different body types and preferences. This can lead to the development of unique and tailored fashion products.

Reducing Wastage: Proper pattern making techniques can help reduce fabric wastage during the production process. By minimizing fabric waste, fashion companies can become more sustainable and environmentally friendly.

Cost Savings: Learning efficient pattern making techniques can lead to cost savings in the production of garments. When patterns are well-optimized, it can reduce the amount of material and labor required to create each piece.

Improved Fit and Customer Satisfaction: Garments that fit well and are comfortable to wear result in higher customer satisfaction. It can help fashion brands deliver products that meet customer expectations, leading to repeat business and positive reviews.

Cross-disciplinary Applications: The knowledge of garment sizing and pattern making isn't limited to garment/fashion design. It can be applied in various related fields such as costume design, textiles, and even the medical industry for creating customized clothing for patients.

Accessibility and Global Impact: Books are accessible resources that can be distributed globally, making knowledge about garment sizing and pattern making accessible to a wide audience, regardless of location.

Industry Advancements: As fashion and textile technology continue to advance, books in this field can contribute to innovations in sizing and pattern making techniques, such as the integration of 3D modeling, artificial intelligence, and digital design tools.

In conclusion, this is a comprehensive book on garment sizing and pattern making, which has a substantial impact on the fashion and clothing industry, from education and skill development to improved product quality, standardization, and sustainability. It plays a vital role in shaping the future of fashion design and production.

Contents

Chapter 1
An Introduction to Anthropometry

Muhammad Umar

Abstract This chapter goes into the intriguing field of anthropometry, which is the study of measuring and comprehending the human body's parameters, dimensions, and variances. It takes a historical trip, delving into the origins of measurements of anthropometry in past civilizations such as Egypt and China, where measurements of the body influenced everything from clothing design to monument construction. The chapter then examines the development of systematic research throughout the Renaissance, when painters such as Leonardo da Vinci used anthropometric ideas in their works. Following that, it investigates the function of anthropometry throughout the Age of Exploration, when scholars such as Buffon undertook groundbreaking studies analyzing body measurements across numerous populations. The chapter addresses the incorrect classifications of races according to cranial measurements that arose around this time. Moving on, it explores the twentieth century's emphasis on standardization and professionalism, with people like Alphonse Bertillon and the formation of organizations such as ISAK playing important roles. Finally, the chapter discusses the different uses of anthropometry in domains like ergonomics healthcare, sports science, and anthropological forensics. It finishes by underlining anthropometry's obstacles and limitations, as well as pointing to the future, where technological developments along with collaboration with other fields hold enormous possibilities for this ever-changing science. Understanding anthropometry's historical history and its many applications gives us a deeper appreciation for the need to accommodate human variations when creating products, services, and settings for a worldwide population.

Keywords Human body · Size · Measurements · Pattern · Scanning · Anthropometry

M. Umar (✉)
Department of Materials, The University of Manchester, Manchester, UK
e-mail: muhammad.umar@manchester.ac.uk

M. Q. Khan et al. (eds.), *Garment Sizing and Pattern Making*, SDGs and Textiles,
https://doi.org/10.1007/978-981-97-7683-2_1

Introduction to Anthropometry

The study is anthropometry—there is a science of the measurement of the human body. Many sizes, shapes, and differences are seen here. An interesting topic is investigated in this chapter. This is not only people's height and weight, but hundreds of details about their bodies, which are looked at here carefully. This science side helps to make what fits well. Costumes are fitted excellently due to dimensions. It is possible to make workplaces for normal human movements. These cases study anthropometry. An example is when people work for an item: their environment suits them perfectly; it was not created strangely. It was because of the consideration of many sizes of human beings. This science looks at the body finely. Indeed, here it studies the sides, sizes, and differences. All these learned from all the people's groups.

Anthropometry is not simply a case of one ruler fits all. In fact, it is more than a comprehensive compilation of checks and measures. It uses an array of methods to identify even the most minute details of the human body. Whether it means gauging your fat stores using calipers, unraveling tape measures around your body, or fashioning digital holograms through 3D scanners. It is far from limited to the four walls of a laboratory. The picture of anthropometry does not end there. In certain respects, its findings are part of the framework of daily existence. The theater of anthropometry can be seen from an excellent ergonomic seating to the exact sizing of favorite wear which is very vital in making favorite commodities safe and convenient for a wide spectrum of users. Those who participate in sports make use of anthropometric data to improve their performance, in that time anthropologists employ anthropometric data while they are working on the mysteries of human evolution and adaptation. The medical profession itself uses these parameters for height assessment, growing monitoring, nutritional assessment, and even in designing prosthetics. The field of anthropometry is no different from a trip of knowledge: you travel not in time nor space, but in understanding. It enables us to ponder the great many differences among human bodies despite being tied with common genetic, ethnic, and environmental fibers. When we have a shared understanding of this variation, then we can leverage it to create a world where human difference is celebrated, and where the human needs of all individuals are met [1].

And yet, the applications for anthropometry are many and varied. This subject is challenging in terms of ethics too: data privacy, informed consent, time, resources, and appropriate interpretation of data in diverse populations all present significant limitations. Addressing these challenges on the road to the future of anthropometry will require a holistic approach in a world of increased technological applications. Examples of possible advancement include the incorporation of wearable sensors and artificial intelligence to help guide future data capture and quantification of the body. The future is exciting, but the field of anthropometry must keep pace to ensure that a future that is human, inclusive, safe, and designed for the glorious diversity of the human body continues to mind the gap.

Definition and Scope

Measurements

Anthropometry, the technological know-how of information human frame length, form, and version, is based on various arsenal of dimension strategies. These techniques flow beyond the easy notion of peak and weight, imparting a nuanced image of our bodily characteristics. Let's embark on an adventure to discover the captivating world of anthropometric measurements as mentioned in Fig. 1.1 [2].

Linear Measurements

Imagine a ruler precisely positioned between anatomical landmarks at the body. This fundamental technique, called linear measurement, bureaucracy is the cornerstone of anthropometry. Here are a few key linear measurements that unveil diverse components of the human form [3]:

Fig. 1.1 Anthropometric measurements

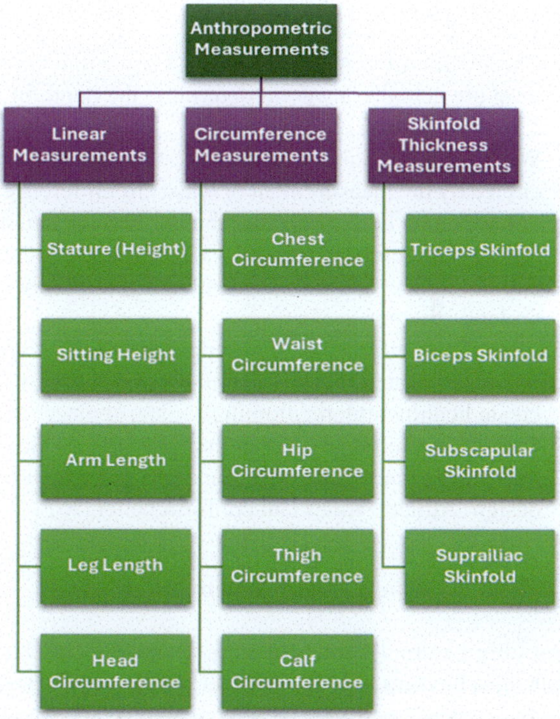

- Stature (Height): The conventional measurement from head to toe, stature gives a basic expertise of average body size.
- Sitting Height: Measured from the pinnacle of the pinnacle to the seat floor while sitting, this measurement is vital for designing ergonomic chairs and workstations that sell the right posture.
- Arm Length: The distance between the top of the shoulder (acromion) and the top of the center finger (stylion) reveals information approximately reach and top body proportions, treasured for product design and ergonomics.
- Leg Length: Measured from the hip bone (extra trochanter) to the ankle bone (medial malleolus), leg duration performs an extensive function in athletic performance, ergonomics, and garment layout.
- Head Circumference: Encircling the top with a tape measure gives insights into brain improvement, especially in infants and children.

Circumference Measurements

While the last one gives linear distance between the points, the second one concentrates on the girth whose calculation is done around the body part. Imagine a tape measure comfortably wrapped around different body parts to capture these valuable data points [3]:

- Chest Circumference: It is widely known that body measurements near the chest level are actually a way to determine the depth of body composition, potentially revealing health risks.
- Waist Circumference: The narrowest portion of the torso just above the hip bones is circled by the measurement of waist circumference, which is one of the indicators of how much body fat is distributed and correlated with health risks.
- Hip Circumference: Explored in most areas of the body and the buttocks, hip circumference is a tool for evaluating the body's proportion, shape as well as skeletal structure.
- Thigh Circumference: I am supposed to take it around the mid-thigh in order to obtain the exact measurement that can be used to determine body mass index and create healthy athletic clothing.
- Calf Circumference: Circumference of calf is measured with it being the point of the thickest calf, but performance in sports can be tracked using this figure for design. Athletic shoes can conform better.

Skinfold Thickness Measurements

Not only external anthropometric measurements but also internal anthropometric data must be measured. The caliper test, though superficial, allows for more in-depth analysis since it measures the subcutaneous fat layer at a specific position of the body. There are specific instruments which we call the skinfold calipers that are used for skinfold measurement. Here are some common skinfold thickness measurements:

- Triceps Skinfold: Located on the back of the arm, this is the percent directly related to the general body fatness. It is easy to use and measure.
- Biceps Skinfold: The first measurement of the skinfold may be taken on the upper arm in front since the obtained figure can be used as part of the indications for the body fat distribution.
- Subscapular Skinfold: We may check under the shoulder blade for the amount of fat stored indicates the upper body fat percentage as well.
- Suprailiac Skinfold: This criterion refers to the fact of having the hip girth or the lower abdomen area being hung over.

Segmental Measurements

Often cases are when either a particular segment of a body draws the attention or the whole body at once. Segmental measurements concentrate on the length of individual body sections, providing valuable information about body proportions: Segmental measurements concentrate on the length of individual body sections, providing valuable information about body proportions [4]:

- Arm Length (Upper Arm + Forearm): Among the measurements that are being used, the one for the upper part of the body breaks down into the length of the humerus (upper arm) and the direction of the radius and ulna (forearm), which allows for more details to be supplied concerning to upper body proportions.
- Leg Length (Thigh + Lower Leg): The same leg length is enveloped by this technique as well by the measuring of the length of the thigh (femur) and lower leg (tibia and fibula).
- Torso Length: It allows the measurement of the torso length from the neck base to the top of the hip bones; this length informs us about the upper body proportions comparing them to the lower body ones.

Advanced Techniques

Technological advancements have introduced new methods for anthropometric data collection, offering greater efficiency and detail:

- 3D Scanning: This technology scans the entire body-related features in a digital format, and then this data is analyzed to achieve the exact measurement of body curves and volumes. The 3D scanning technology is increasingly being used in anthropometry studies to construct an all-round profile of body size and geometry.
- Photogrammetry: Such method is the 3D body model generating procedure which includes multiple side photos of the body.

Applications

Multiple branches make anthropometry more than just gathering data as mentioned in Fig. 1.2. It aims to:

Understand human variation: It is amazing how anthropometry reveals the enormous variations in human figures compared to different populations (e.g., Europeans, Asians, Africans), different age groups (child, adult, elderly), genders (male, female), and ethnicities. This knowledge is a fundamental condition for designing items and environments that are comprehensive and serve an increased number of users.

Inform design and product development: Knowledge gained from anthropometric data enables designers, engineers, and developers of products, workplaces, and environments to ensure that they are safe, comfortable, and accessible to a broad spectrum of users. This way, such as chairs of the airplanes, car panels, and workstation settings takes into account the proportions of people in order to avoid discomfort or potential injuries.

Advanced scientific research: Anthropometry(s) offers the most important information for several scientific subjects being used in:

Human evolution: The molecular and genetic data analysis yields the various development trends and anatomical peculiarities of different humans and their behavior in different habitats over the years.

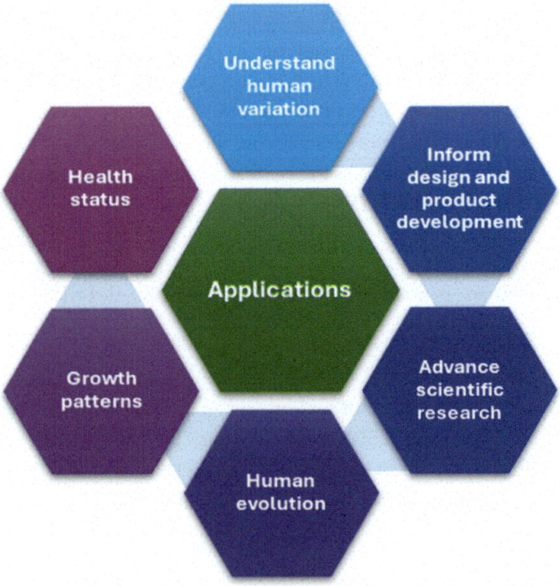

Fig. 1.2 Applications for anthropometric measurements

Growth patterns: Anthropometric values of children and adolescents can be utilized for tracking growth and development patterns, finding potential nutritional deficiency cases, and discovering health status.
Health status: The body mass index, which is determined from the height and weight measurements is believed to be the most used indicator of the possible health challenges associated with being underweight or obese.
Athletic performance: Anthropometric data is used to evaluate athletes' physical capabilities and predict whether their inherited attributes can offer them the possibility of doing well in particular sports. The training program's execution quality is enhanced by body composition (muscle mass to fat ratio) and proportions analysis as well as it emphasizes personalization.

Historical Background

It was anthropology that first witnessed the systematic studies of human body dimensions in the eighteenth century although the body shapes and measurements can be found in ancient civilizations [5]. Here's a more detailed look at some key milestones:

Very Early Civilizations (3000 BC–1500 AD)

Sizing the body of the Egyptians for everything from clothes design to proportioning people's measurements was a vital task when making garments in order to accommodate clothes that would fit on the pharaohs and other members of the society. The Indus Valley population probably had a high-degree knowledge of anthropometry that they applied in town planning and sculpturing with single structures and statues made proportionate to the human size. In China, the ancient medical system used morphometrical principles in pulse diagnosis and acupoint location, where a knowledge of body proportions and pressure points was the important guides.

Renaissance Was the Period Around (1500–1800 AD)

Artists such as Leonardo da Vinci could be revered since they made a nationwide pursuit for studying the body and proportions. They found a humanistic approach present in their work. The most popular of his works would remained the Vitruvian Man drawing representing, via mathematical ratios, the perfect structure of the human being based on body measurements. Among the scholars much related to ergonomics was Georgius Agricola, a German physician and mineralogist, who primarily used anthropometry for Explanation: Anthropometry was introduced by Georgius Agricola, a German physician and mineralogist, who specifically used it for ergonomics.

He explored using human body measurements and making tools that fit the humans they were built for rather than each individual having to adapt to them; this was to relieve tension between the body and the tool and therefore enhance working speed.

The Eighteenth Century Marked the Start of the Age of Exploration (1800 AD–1900 AD)

This is what holds the history of the humans throughout this period when explorers meet people from different corners of the world which features diversity. The naturalist and anthropologist Georges-Louis Leclerc Comte de Buffon even measured body sizes among the diverse populations from which he sought to explain human variation, thus setting the academic stage for us to record patterns that we now know today. A fellow German anthropologist, called Johann Friedrich Blumenbach in an attempt to classify races based on the skull dimension. Nevertheless, it has to be acknowledged that the system was created on the base of errors and is now rejected after the issues with the racial ideology came out.

Continuous Vanquishing of the Unknown: Part I (1900 AD–Present)

We can observe the transformation of the anthropometry from twentieth century: from the basis of standardized technique and to the completely professional discipline. This prevented discrepancies in the data, thus giving enhancements in consistency, accuracy, and comparability of the data findings regardless of the location and time. Here are some key developments:

Alphonse Bertillon (1853–1914): A French criminalist, Bertillon was working on a system (that was afterward displaced by fingerprinting) based on personal anthropometric data (like the length of the head and the length of the foot and the span of the arm). The system was designed for identification of criminals. In the long run, upon the emergence of fingerprint analysis, the anthropometric method might have been replaced, but it remains one of the earliest and meaningful techniques in standardized hands measurement.

International Society for the Advancement of Kinanthropometry (ISAK): In 1971, ISAK came into existence under the umbrella of association that aims to create research, education, and a framework for the kinanthropometric standards, which is a branch of anthropometry that concentrates on the aspects of human structure, body movements, and overall performance. ISAK provides the guidelines to minimize human error with measurement protocols, terminology, and collection methods, all of which make the anthropometric research worldwide to be efficient.

Importance in Various Fields

Anthropometry ranges from a wide selection of usage and applied to various fields as an impacting factor. It's worth taking a closer look at some of the most important applications as you will see the following.

Ergonomics

Ergonomics is a field of science that aims to create workplaces, products, and machines that not only are safe and comfortable but also adaptable to all types of workers regardless of their build and abilities. Anthropometry which provides data about the dimensions of a person, postures, and abilities has an important role on ergonomics [6]. This data is used to:

Design workstations: Working on anthropometric data like reaching the height of elbows, distances, and legroom, ergonomists can develop stations that are meant to eliminate musculoskeletal stretches and enhance pronunciation of posture, and thus the chances for work-related injuries get reduced.

Develop comfortable products: Thanks to anthropometric data, designers can model chairs, desks, hand tools, or all kinds of household appliances in the shape of human body, human-friendly, and convenient for diverse customers.

Ensure accessibility: Observing anthropometric data for design purposes will guarantee the suitability of both products and overall environment for those suffering from disabilities. Take for instance, widening doorways, hallways, and kitchen countertops that will use appropriate building plans to accommodate the disabled and their aids.

Anthropology: Anthropometry continues to retain its relevance and importance among researchers in anthropology who are interested in human evolution, the different types of variation and adaptation across isolated, geographically diverse populations and cultures. Body measurements provide valuable insights into:

Migration patterns: Through assessing human data across different populations and by the analysis of skeletal remains, anthropologists can gain a better understanding of historical migration patterns that humans moved from one place to another and how human populations or groups changed to different environment throughout time.

Environmental influences: Bigger people can be caused by factors like diet and environment, such as the climate. Anthropological research with skeletal remains can contribute to the accumulation of information on how people bodies have altered while adapting to the change of environment.

Basic Concepts and Terminology

Knowing many basic concepts and terms is sure to guide the attempt into this field, which is anthropometry. Here's a breakdown of key elements:

Body Segments: The human body is like a giant puzzle to be put together according to the anthropometric limiting time. With these segments yielding research results, researchers can be comparative and take measurements from different populations. Common body segments as mentioned in Fig. 1.3 include:

- Head
- Neck
- Breast
- Part of the upper limbs like arms, hands, and shoulders
- Lowers Limbs (It is thighs, feet, and legs).

- Head
- Neck
- Breast
- Part of the upper limbs like arms, hands, and shoulders
- Lowers Limbs (It is Thighs, Feet, and Legs)

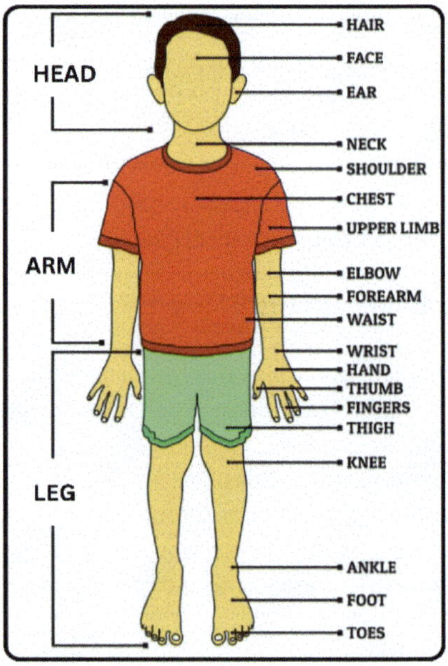

Fig. 1.3 Common body segments

Landmarks: Particular anatomic body prompts measure specific points. These are the most tough external points when measurements are to be performed. Thanks to some of these landmarks, they can be easily spotted and their size and location aren't going to be too different between individuals which is important for the analysis to be consistent. Examples of anatomical landmarks used in anthropometry include:

- Head: C (Vertex that is the top point of the head) and B (Glabella that is the bony ridge above the eyebrow).
- Torso: Jugular notch (extra deep hollow within the base of the throat), Xiphoid process (end portion of the sternum)
- Upper Limbs: A1 The distance between acromion (bony prominence at the shoulder) and Stylo (bony prominence at the wrist) define the length of the arm.
- Lower Limbs: Sudoku (Hip bone protrusion) and Medial malleoli (inner ankle projection)

Standardization and Measurement Techniques

Standardization provides the assurance of unified measurements and reduces the miscalculations by anthropometric data collecting. Standardization, therefore, stands out as a core element in the successful formulation of a refined product development roadmap.

Defined Procedures: Normalized protocols which provide precise directions on how to acquire each of the measurements. These protocols include details such as:

Body position: Time and frequency of intervention, whether the participant should be standing, sitting, or lying down.

Instrument placement: Whether it is a local spot on the human body or it is a recording device that must be kept nearby.

Reading technique: What is the way to read the measurement from the instrument and to record them accurately.

Calibrated Instruments: Anthropometry devices, e.g., stadiometers (height) are to be rechecked and sorted out on a regular basis to ensure they deliver right results. The process of calibration entails the alignment of the imputed values by a known standard for confirming that the instrument is correct in its reading.

Trained Personnel: Scientists who have been specially trained are able to measure individuals using the proper methods minimizing the involuntary errors caused by prejudice. They are practicing instruments, recording data, and using them in the correct way and positions.

Accuracy and Precision Issues Are Created by Innovative Medical Technologies

Accuracy and precision are two crucial aspects of anthropometric measurements as explained in Fig. 1.4

Accuracy: In this case, the term deals with how close a measurement is to the precise value of the body dimension that is predicted. Data accuracy can be determined by multiple variables as good or bad such as calibration of instrument, the method technique of the practitioner and cooperation of the participant. Another instance of this inaccuracy would be readouts of measures of height from an incorrectly calibrated stadiometer.

Precision: References randomness of a measurement. Use our AI to write for you about: Small businesses and the local economy: challenges and opportunities. It can be done either by the same or multiple anthropometrists using the same method of measurement and assessment to provide the standard for precision. The term "high precision" means that there is no variation from one measurement to another and that the measurements are immune to external disturbances.

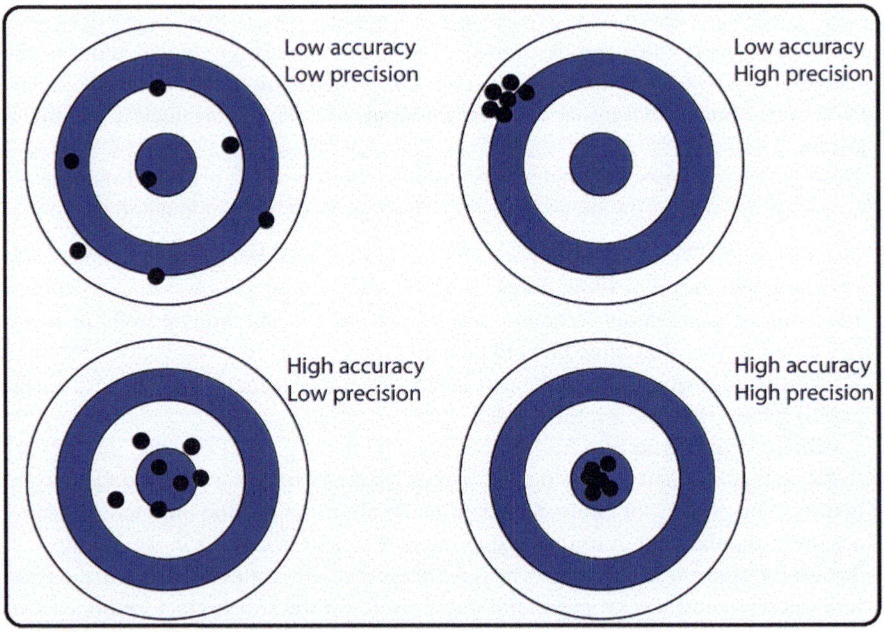

Fig. 1.4 Example of accuracy and precision

Applications of Anthropometry

Here's a deeper dive into some of the key applications of anthropometry across various disciplines.

Product Design and Ergonomics

Safety Considerations: To ensure product safety, anthropometric data are of pivotal importance for manufacturers. In the case of car seatbelt design and airbag deployment systems, human data related to anthropometry is taken into account to provide optimum protection for passengers of different proportions when they experience a crash.

Accessibility Standards: Techniques in anthropometry provides the essential elements which are the basis and development of accessibility standards for products and environments. Entrances, corridors, and public transport systems are created based on the relation to the size of the person so that people on wheelchairs, walkers, or other accessibility devices can use them easily.

Biomedical Applications

Nutritional Assessment: Anthropometric measurements which came as weight, height, and skinfold thickness are the form of information that healthcare professionals make use of to evaluate nutritional status as well as to identify any deficits or potentials weight-related health hazards.

Growth Monitoring: Following the growth and the development of children, on the other hand, over a period of time is an important aspect of healthcare professionals because it helps them to monitor the growth and the development patterns, in addition to identifying potential issues such as growth delays.

Prosthetics and Orthotics: Customized partial prosthesis or orthosis making depends on the proper data collection with custom fit, the support balance, and performance for amputees or movement-restricted individuals [7].

Sports Science and Athletic Performance Optimization

Injury Prevention: Physiologists and trainers can identify muscle imbalances and scrutinizing body proportions with a set of anthropometric measurements may be used to design programs that help prevent injuries. For instance, a runner who has concordance in leg length can benefit from some strengthening exercises so as to improve the muscle that holds the body stabilize and the proper running form.

Equipment Design: The reason is that sports gear like helmet, protective gear, uniform and so on has been based on a model of the size and body measurements to reflect on the size and the shoes of different athletes.

Challenges and Limitations

Despite its extensive applications, anthropometry faces certain challenges that need to be considered:

Ethical Considerations: Gaining valid participant consent and maintaining data confidentiality must be the primary ethical components of anthropometric studies. Data collectors need to be very open about the process of data collection, use, and storage so that the public trust is preserved.
Practical Constraints: Taking anthropometric measurements is particularly time-intense and needs specimen measuring devices and experts with the proper skills. This can be the hardest thing to handle as the community may lack the resources and professionals who are well experienced enough to handle the situation.

Interpretation Issues

Anthropometric data needs to be interpreted with consideration for several factors:

Variability within Populations: Body size and shape may show substantial variety among populations dependent on the genetic makeup, ethnicity and socioeconomic class of the people. Scientists should keep this named difference in mind when they are looking at data and avoid fashioning conclusions in the most general form.
Outliers: Both those with height, weight, or other measurements beyond typical range for a certain population and so-called individuals with height, weight, or other measurements beyond typical range for a certain population should be especially studied. Medically known underlying conditions, or genetic reasons, can distinguish these outliers from the usual population patterns.

Future Directions in Anthropometry

The field of anthropometry is constantly evolving, driven by advancements in technology and the need to address emerging challenges.

Advances in Technology

Wearable Sensors: Gadgets such as fitness bands and smartwatches can take data continuously on motion characteristics, posture, and levels of activity. These datasets enable incorporation of dynamic human movements into the anthropometric measurements data which would describe whole human activity, and not only its reflection at a specific time point.

Artificial Intelligence (AI): AI algorithms can analyze huge amounts of anthropic listening data to spot trends and patterns in it. This increased comprehension may result in an improved knowledge about individuals' health, performance, and design.

Association with Others Subject Areas

Biometrics: Synthesizing the information of height, weight with metric indications of fingerprints and face recognition presents a more secure and comfortable mechanism of authentication. For instance, the data on anthropometry like hand geometry could be authenticated with the fingerprint scanner to ensure higher security.

Robotics: Human data on anthropometric indicators should be incorporated into the design of robots to make them capable of safely communicating and teamwork with people regardless of their sizes and shapes. This is expedient for the designing of robots that will be used in industries, homes, etc. and at the healthcare settings.

Focus on the Duly Existing Social, Political, Economic, and Environmental Challenges

Aging Populations: While lifespans are a key driver to increase the need of beneficial devices and equipment like assistive technologies and facilities, which mainly serves the purpose of supporting older adults. To illustrate, senior living facilities and public transport systems building ramps and grab bars become the key factor as well as furniture higher suitable for older adults having the possibilities of mobility issues.

Virtual Environments: Learning anthropometry (the science of measuring the body) will be an important prerequisite for developing VR or AR content that is comfortable and engaging for people having diverse physiques. To afford a VR videogame story and AR head/eyewear is to develop devices that can accommodate any type of a shape and a size of the head in order to offer users comfort.

Conclusion

The Illustrious Discipline of Anthropometry Always Remains True: It Is Through the Synergy of the Human Body and Its World Around It that Science Manages to Extend Its Field. Through examining the human body size, shape, and variability, it gives us the accessible data that guide the theories used in different spheres by many scientists. Through the development of ergonomic and comfortable products to the creation of strategies for athletes optimization and the comprehension of issues related to health, anthropometry assumes a major role in the shaping of the globe for different human groups.

Technology and invention into the future will go on changing and fresh issues raising therefore, anthropometry will always be a significant tool. The gains made in wearable sensors, artificial intelligence, and integration with other sectors like biometrics and robotics exhibit high potential for the advancement of anthropometry in the future. Anthropology that adopt these developments and deal with the ethical and practical issues can continue to be among key tools that help us understand the human shape and its universality.

Key Points Summary

- Anthropometry is a discipline that deals with the analysis and investigation of bodily dimensions, proportions, and dissimilarities that characterize our body.
- Methodology employs diversity of techniques for measurements, among which, there are linear measurements, circumference measurements, skin assess thickness measurements, and segmental measurements.
- Development of technologies such as 3D scanning and photogrammetry is growing at a rapid speed, as they being used to monitor information.
- Anthropometry finds its use in multiple disciplines, such as ergonomics, anthropology, medicine, human factors, and forensic anthropology.
- One of the main challenges is to ethically, practically tolerate data interpretation limitations.
- Due to the technological breakthroughs as well as the achievements in the interdisciplinary integration there are a lot of varying options for the future anthropometry.

Significance of Anthropometry

Much more than just measuring the physical human body with the help of anthropometry, the importance of this science can be emphasized. It is unbelievable how these findings contribute to every aspect of our lives, from designing the objects and models that integrate the foundations of marvelous human beings. Here's a deeper look at the profound impact of anthropometry:

Understanding Human Diversity

Another pivotal role of anthropometry is to explain the outstanding diversity which is the human population. As the world faces the threat of widespread food insecurity and hunger, the role of sustainable agriculture has become increasingly vital. Through the use of body size and shape and proportions of each ethnicities, genders, and geographic regions, anthropometry gives us a bigger picture that connects us with the whole world. This knowledge is essential for:

Inclusive Design: For an example, anthropometrical data helps all sort of things, starting from clothes to spacecraft cabins and places to be, to have a good fit for more people feeling at ease while wearing or using them. An example is designing doorways and hallways based on anthropometric data to allow large users of wheelchairs and walkers to change the width of these openings to those sizes that will accommodate them. The same designers of clothing outfit sizes also use the anthropometric data to design body shapes that accommodate different body types.

Combating Bias: Knowing the diversity of the humankind helps to break stereotypes and biases with regard to physicality. Anthropometry is clear evidence that such a spectrum of what is considered normal to exist within our species is a wide one, consequently it brings up issues of inclusivity and respect for individual differences.

3D printing technology enables real-life replication, making it an ideal tool for product designing and development

This discipline, while it may not be as mesmerizing as others, is crucial to the quality and design of many everyday articles, from the chair you seat on to the clothes you wear. It ensures these products are:

Safe and Efficient: Through the knowledge about dimensions of people's bodies, capabilities, and also their limitations, designers can create products that enhance safety and, at the same time, befitting success. For instance, cars' adjustable seats and control panels are initially developed through anthropometric data during the driver's construction, which aims to support drivers of different sizes and make sure they can comfortably and safely enjoy a good driving experience. Apart from that, the device and tool designs take account of anthropometric data to avoid them from being too large or small and thereby, causing unnecessary musculoskeletal strain.

Comfortable and Usable: Imagine having a mouse that is specially designed for your hand, so only the movements are convenient to you, or a pair of headphones that always suit your head and ears perfectly. Body measurements assist in tailoring designers' products to fit the human form precisely and they enable a full range of users to use them comfortably well. For example, when designing furniture components, such as arms, seat depth, and leg space, a designer keeps in mind anthropometric, reliable data on different heights and body types to make sure the furniture will fit and be comfortable for all.

A Foundation for the Future

As technology continues to evolve, anthropometry will undoubtedly play a key role in shaping the future:

Personalization: Technology development including the emergence of wearable sensors and artificial intelligence is the backbone of customized individual applications and completely revolutionizes the medical technology field of use. The capability of that technology is the ability to amass anthropometric information by professionals and use it to design individual routines, for example, workout plans, or customized recommendations, for instance, product suggestions.

Virtual and Augmented Reality: With virtual and augmented reality becoming more known, anthropometric data will be the prerequisite for designing comfortable and immersive interfaces for a diversified user so the size can fit in. VR headsets and AR glasses will need to be engineered such that can fit in any head and in addition that should be as comfortable as possible for the users for instance in head shape and size.

Robotics and Automation: The assimilation of anthropometric data into robotics will be the essence of the safety of robots in human interactions. Robots will be of different sizes and forms. This is the very condition for creating a robot that will be possible to introduce into workplaces, houses, and healthcare facilities in a natural way.

References

1. J. Wang, J.C. Thornton, S. Kolesnik, R.N. Pierson, Anthropometry in body composition: an overview. Ann. N. Y. Acad. Sci. **904**(1), 317–326 (2000). https://doi.org/10.1111/j.1749-6632.2000.tb06474.x
2. G. Nadadur, M.B. Parkinson, The role of anthropometry in designing for sustainability. Ergonomics **56**(3), 422–439 (2013). https://doi.org/10.1080/00140139.2012.718801
3. M.A. Minetto et al., Digital anthropometry for body circumference measurements: European phenotypic variations throughout the decades. J. Pers. Med. **12**(6), 906 (2022). https://doi.org/10.3390/jpm12060906
4. C. Raja, R. Singh, H. Bharadwaj, Anthropometric determination of body volume, body density and segmental volume in adult Indian women. Ann. Hum. Biol. **4**(6), 565–575 (1977). https://doi.org/10.1080/03014467700002561
5. J. Baten, M. Blum, Growing tall but unequal: new findings and new background evidence on anthropometric welfare in 156 countries, 1810–1989. Econ. Hist. Dev. Regions **27**(sup1), S66–S85 (2012). https://doi.org/10.1080/20780389.2012.657489

6. I. Dianat, J. Molenbroek, H.I. Castellucci, A review of the methodology and applications of anthropometry in ergonomics and product design. Ergonomics **61**(12), 1696–1720 (2018). https://doi.org/10.1080/00140139.2018.1502817
7. N. Utkualp, I. Ercan, Anthropometric measurements usage in medical sciences. Biomed. Res. Int. **2015**, 1–7 (2015). https://doi.org/10.1155/2015/404261

Chapter 2
History of Sizing System

Phan Duy Nam and Muhammad Abbas Haider Alvi

Abstract This chapter explores the development of clothing sizing methods over time, from the days of custom tailoring to the current era of mass-produced apparel. It looks into how the Industrial Revolution, trade laws, guilds, and royal decrees all influenced the creation of standardized size. The chapter examines the difficulties and constraints associated with early size attempts, as well as the development of contemporary sizing tables and charts in the twentieth century. In addition, it looks at how the World War II and post-war commercial culture affected sizing conventions and critically assesses modern sizing schemes, calling attention to problems like "vanity sizing" and inclusion constraints. The chapter concludes with a look at how size may develop in future, highlighting developments like as 3D scanning technology, the emergence of adaptive clothing elements, and the possibility of more environmentally friendly manufacturing methods. Through an appreciation of the historical background and constraints of existing systems, the chapter opens up possibilities for future developments that guarantee customized, well-fitting apparel experiences for a multicultural world population.

Keywords Garments sizing · Sizing system · Pattern · Marker · Sizing history

Introduction

Human bodies vary widely in size and shape. There are essentially an endless amount of different body shapes that may be created from this constant change. It is necessary to translate continuous, mostly normally distributed variation into a few body types, known to customers as "sizes," when dealing with practical issues of recognizing

P. D. Nam (✉)
Faculty of Textile—Leather and Fashion, School of Materials Science and Engineering, Hanoi University of Science and Technology, Hanoi, Vietnam
e-mail: nam.phanduy@hust.edu.vn

M. A. H. Alvi
Department of Textile Engineering, School of Engineering and Technology, National Textile University Faisalabad, Faisalabad, Pakistan

M. Q. Khan et al. (eds.), *Garment Sizing and Pattern Making*, SDGs and Textiles,
https://doi.org/10.1007/978-981-97-7683-2_2

human body shapes and sizes, such as in the production of apparel. This is necessary for both the orderly production of garments and the ease with which customers can choose their desired items. The human body's metric characterization necessitates collecting numerous measurements because it is a geometrically complex shape that also fluctuates allometrically in size of various body dimensions. Size changes result in altered proportions. The shape information that is present in their interactions is not fully recorded, and these are typically interpreted as circumferences or point-to-point lengths. Since these interactions are intricate, it is not possible to simply reduce a significant amount of data by selecting one or a small number of "key dimensions" or their ratios. This explains why despite frequent criticism in the media, clothing sizing systems that rely solely on a few measurements, such as bust circumference, do not satisfy customers. Adding one or two more body dimensions—such as stature and hip circumference—to the sizing method does not significantly improve the problem. It is not realistic to add more body measurements to an industrial sizing system since this would force manufacturers to create an excessive number of "sizes" and make it difficult for customers to have to memorize many body dimensions in order to try to match them all with a specific item in a retail store.

The rise of mass production and ready-to-wear (RTW) led to the creation of sizing systems. The body sizing system and the garment sizing system are the two different kinds of sizing systems. While a garment sizing system starts with a body sizing system and is evaluated on prototypes for fit, a body sizing system is established using close-to-body measurements. Standard body measuring methods were originally studied by academics in the first half of the twentieth century, and their popularity peaked in the latter half of the same century. Efficient sizing systems are advantageous to producers and customers. Researchers have discovered that customers were not happy with how the clothing fit, though. The fact that companies have been using out-of-date grade rules is one of the causes of this problem. Another factor is that RTW firms only offer a few selection of sizes, particularly for the plus-size market. Creating a modernized, more adaptable sizing method is crucial [1].

The trend in the apparel sector is mass customization. One of the objectives of mass customization is to provide a better fit quickly and at a reasonable cost. Mass customizing sizes are one way to address this. Custom fit patterns can be created in three different ways: (a) flattening three dimensions into two dimensions; (b) automatically creating two dimensions; and (c) modifying from standard sizes. Having an effective technique for generating standard sizes based on anthropometric data is essential for the last choice. Three-dimensional scanning may no longer be limited by space thanks to Microsoft's development of Kinect technology, a range of motion-sensing input devices that can estimate distance and create 3D models. Body measuring data may become more widely available in the future than they are now, which may present both benefits and difficulties. In the near future, the garment industry will need to decide how to fully utilize and evaluate the data. There is ample evidence of consumer unhappiness with fit, and 3D body scanning tools have made measurements easier to get. This begs the question of how a sizing system could be improved to better suit the needs of customers. Thus, the purpose of this study was to investigate and devise a procedure for the development of a sizing system that incorporates more recent anthropometric data.

Definition of Sizing System

A sizing system is a set of uniform measurements and categories that are used to identify the right size for a variety of things, including manufactured goods, apparel, shoes, and equipment. These systems are made to make it easier for goods to be produced, distributed, and consumed consistently and uniformly across many marketplaces and industries.

Typical essential elements of a sizing system are as follows:

Measurements: Depending on the product being sized, specific dimensions or metrics—such as length, width, height, circumference, or other pertinent measures—are employed to define its size.

Labels or Categories: Product classification techniques that divide items into several size ranges or groupings. These groupings could be based on other identities, letter codes, numerical sizes, or descriptive phrases like small, medium, or giant.

Standards and Guidelines: Specified guidelines or standards, frequently derived from anthropometric data, industry standards, consumer preferences, and other pertinent aspects, are used to determine proper sizing.

Sizing Charts and Guides: Resources or instruments offer direction on choosing the appropriate size based on personal measurements, body types, or other pertinent aspects. Retailers, manufacturers, or trade associations may offer these materials.

Sizing systems have changed over time in response to shifting industrial methods, customer preferences, social mores, and scientific developments. Standardized size was developed to address practical concerns like fit, comfort, and safety for end users, and it has been crucial in facilitating trade, business, and communication in international markets.

Early Developments in Sizing Systems

Since the Industrial Revolution, there has been apparel that is ready to wear. In the past, clothing was custom-fitted for each wearer by professional dressmakers or tailors, family members, or the wearer themselves. Between 1820 and 1838, the proportionate dressmakers' systems were the first grading schemes in use. They assessed the other pattern dimensions equally based on a single body measurement, such as the bust measure. These were not based on empirical anthropometric research, but rather on the experience of tailors. The eighteenth-century wars led to growing armies and mass uniform manufacture, which necessitated the development of a systematized size grading system and the collection of statistical data regarding men's body measurements. The first extensive scientific investigation of women's body measurements took place in the USA between 1939 and 1940. British sizing during the first part of the twentieth century was commonly divided into 2-inch increments for the bust, waist, or hip size. Size identification codes, such as SW (small woman), W (women's size), OS (outsize), and XOS (X-outsize), were also employed. A 36–38 inch breast would fit a typical woman's size W. Some British manufacturers began using American number coding systems; women's sizes 10, 12, and 14 are examples. By the latter half of the twentieth century, efforts were made in both the USA and Europe to create consistent sizing using body measurement surveys and the use of statistical methodologies, as the many size identification systems were already thought to be confusing [2].

In 1953, the British Standards Institution (BSI) released the first scientific size charts, which had a predetermined code sizing pattern that was closely correlated with body measurements. The US Department of Commerce produced CS 215-58 in 1958, which were the first standard garment sizes in the country. The first international standard for clothing size designations, complete with definitions and body measurement protocols, was issued in 1977 (ISO 3635). The development of an international sizing system for apparel began in 1969. A modified version of this standard (EN 13402-1 2001) has been adopted by the European Committee for Standardization, which is currently working on creating a new European size identification system (prEN 13402-4). They've had trouble coming up with a standard size code; it needs to be accurate and informative, without being too complex for consumers from other countries to understand or for the clothing industry to adopt.

Clothes reflect culture, history, and social status in addition to serving as a means of bodily protection. These variables have also influenced how we size clothes, creating an intriguing global patchwork of disparate systems. Let's examine how size systems have changed throughout time in many cultures [2].

Ancient Clothing as a Status Symbol:

- Clothes acted as a status symbol in numerous ancient civilizations. Only kings and nobles were allowed to wear elaborate clothes that came in particular sizes. For instance, broader collars on clothing in Ancient Egypt denoted a greater social standing.
- Regional Variations: In contrast to modern methods of uniform measuring, ancient cultures frequently used descriptive terminology or precise measurements based on bodily parts. When arms are extended, the length between wrists can be used to determine robe size in China.

Customized Stitching and Local Customs

- The Tailor's Artifice: Before clothing was produced in large quantities, it was fashioned to order by talented tailors who depended on their understanding of each customer's body. One of the best examples is the Japanese kimono, for which tailors employ an intricate system of measurements unique to the wearer's posture and ideal fit.

- Methods of Regional Measurement: Many cultures have evolved distinctive methods for measuring clothing. For example, the length of a sari in India is measured using the wearer's arm span (hath) as the base unit.

Colonialism and the Disruption of Tradition

- Western Imposition: As a result of colonial expansion, other nations were forced to adopt Western size standards. This caused a rift between indigenous body types and Western sizing, upending long-standing customs.
- Cultural Opposition and Adjustment: Not every culture acquiesced to enforced structures in a passive manner. Some kept some components of their own customs while incorporating Western size. As a result, hybrid systems reflecting regional preferences were created.

The Rise of Readymade Clothing and Regional Challenges

- Changing Western Systems: The arrival of readymade apparel posed a problem. Western sizing standards have failed to take cultural differences in body type into consideration. For instance, there may be disparities in sizing because individuals of Asian heritage may be smaller in build than the typical Westerner.
- Emergence of Regional Standards: Some areas created their own sizing guidelines that more accurately represented the local populace in order to address these problems. For example, Japan has a different sizing system for clothes sold there.

Modern Dilemmas in a Globalized World

- Vanity Sizing and Worldwide Confusion: The emergence of "vanity sizing" has brought out a worldwide dilemma. In an attempt to appeal to consumers, brands may label clothing as being smaller than the real measurements. International customers become frustrated and confused as a result.
- The Need for Inclusivity: More inclusive size guidelines are gradually being adopted by the fashion industry. This entails providing a greater selection of sizes and accommodating various body shapes.

The Future: Technology and Personalization

- 3D Body Scanning: Technological developments provide encouraging answers. Regardless of location, 3D body scanning may produce an accurate digital model of an individual's body, enabling custom-fit apparel.
- Accepting Diversity: Accepting cultural diversity is essential to the future of scaling systems. This entails creating standards that accommodate a wider range of body forms and are sensitive to regional variances.

Standardization Efforts in Medieval and Renaissance Europe

The search for uniform clothing sizing is a tale of human inventiveness and the changing dynamics of trade, power, and apparel see Fig. 2.1. The age of size charts was genuinely ushered in by the Industrial Revolution, but intriguing if complex, foundational work was done by guilds, royal decrees, and trade restrictions.

Guilds: Beyond Stitches and Secrets

- Knowledge Guilds and the Power of Tradition: Guilds were more than just fancy clubs for artists with talent. They served as knowledge archival tools, transmitting from generation to generation procedures for tailoring, clothing manufacture, and maybe early sizing practices. Guild apprenticeships offered a disciplined learning environment where aspiring tailors refined their abilities in both sewing and the technique of fitting clothes to different body shapes.
- **Standardization Through Shared Vocabulary**: It's possible that guilds have created a uniform vocabulary to describe clothing sizes and physical proportions. This lexicon would consist of terminology exclusive to a given guild or area rather than an all-inclusive size chart. To promote uniformity among its members,

Fig. 2.1 Standardization efforts in medieval and renaissance Europe clothing

a Florence-based guild may utilize terminology like as "palmo" (hand breadth) or "braccio" (forearm length) as benchmarks for measures.

- **The Limitations of Guild Standardization**: The important disclaimer is that this standardization was probably anything from uniform. Guilds were frequently regional organizations with very dissimilar customs. Furthermore, they prioritized craftsmanship and quality. In contrast, the concept of sizing remained more flexible due to individual needs and the ability to customize items. Although a Florentine tailor may utilize "braccio" to determine sleeve length, the precise measurement for a given customer may vary based on preferences or the style of clothing that they choose to wear.

Royal Decrees: Fabric, Fashion, and Power Plays

- **Controlling Luxury and Displaying Opulence**: Royal proclamations were frequently employed as a means of controlling the materials and adornments used in apparel, especially for the nobility. These regulations served as a potent statement of authority and riches in addition to being aesthetically pleasing. Monarchs demonstrated their authority and sway by limiting the usage of particular textiles or decorations to the aristocracy. Deeds may also increase home textile manufacturing by reducing the competition from foreign materials.
- **Indirect Impact on Garment Size**: These decrees may not have explicitly prescribed clothing sizes, but they may have indirectly affected garment proportions. To accept a new type of material with a different breadth, for example, tailors may have created clothing with a more fitting profile in response to a decree limiting the import of a certain sort of wide, flowing fabric.
- **Limited Scope and Patchy Enforcement**: It's crucial to keep in mind that royal edicts mostly affected the attire of the nobility. They most likely had little impact on what the broader public wore. There's also a chance that these directives won't always be enforced consistently. The time period, the place, and the crown's resources all affected how effective it was. A strong monarch in a consolidated kingdom may impose stricter regulations on his subjects than a weak ruler in a dispersed area.

Trade Regulations: A Global Tapestry, Unevenly Stitched

- **Quality Control and Import Restrictions**: A common goal of trade laws was to guarantee the caliber of imported textiles. These rules may outline minimum thread counts, particular weave patterns, or even the fabric's measurements. A regulation might, for instance, mandate that imported silk meet certain width and thread count requirements.
- **The Ripple Effect on Garment Size**: Like royal decrees, these rules might have indirectly affected the size of clothing. Sewing methods and styles may have been modified by tailors to make the best use of the material if a law mandated that imported linen be a certain length. Depending on the kind of fabric utilized, this can have resulted in differences in clothing sizes.
- **Regional Variations and Evolving Trade**: Trade restrictions would have had a very varied effect on sizing. It was mostly dependent on the area and the particular

trade agreements that were in force. Furthermore, as new trade routes opened and political alliances changed, trading patterns changed continuously. Due of these issues, establishing a stable, consistent size system via trade regulations alone proved challenging.

Industrial Revolution and Mass Production

Taking measurements of the human body is the first step toward creating body-fitting clothing. Measuring instruments vary in complexity and low- and high-techness. Providing clothing for "everybody" was one of the objectives of twentieth-century mass production and distribution networks. Was this objective accomplished? In little than a century, Americans transitioned from wearing mostly tailored clothing made by a skilled professional or fitted and sewed at home to wearing clothing that was mass produced and sized by an unidentified group of professionals. If we are to obtain clothing with a suitable fit, we must recapture the unique relationship between the body and clothing, which was lost during this process. In order to do this, we must comprehend the subtleties of body dimensions required for a well-fitting product that satisfies the customer's practical and aesthetic needs at an affordable price range. We are still far from meeting the objective of offering "everybody a fit of the same quality" even in the twenty-first century. Rather, we have disregarded the individual body and assumed that every body will fit into clothes that are the same size. Sadly, this has led to the expectation that a person's physique would conform to a certain norm for clothing rather than creating apparel that fits every. The apparel industry has the potential to offer individually tailored and designed clothing with the emergence of mass customization models. Finding methods for measuring the human shape and creating clothing that fits the body, however, still requires a lot of work. Creating dependable and accurate body measurement techniques is one piece of the puzzle. This paper's goal is to examine and assess both historical and contemporary approaches to body measurement in order to help map out the future paths required to produce clothing for "every body." The methods used for measuring and interpreting the human body are essential to the creation of an outfit pattern. It's not a precise science to size clothes. Although many fields utilize body measures to record health and history or create items like prosthetics and ergonomic chairs, the fit relationship between the body and clothes makes the measurements required for clothing unique. Finding dependable body landmarks that connect the body to the pattern directly is necessary to convert a body measurement into a clothing pattern. The body's natural bends or bony protrusions might serve as identifiable landmarks; soft parts of the body are more difficult to locate. Getting precise measurements in person is essential to getting a good fit. Understanding the objectives of clothing firms, consumer preferences, and the uses and functions of specific items of clothing are other crucial components [3]. Even while we don't really want to go back to the custom tailoring of the 1800s, we won't put up with ill-fitting clothes thanks to new, readily available technologies. The way the clothing business develops in the future will rely on how well we comprehend and take into consideration the different factors that affect how a garment fits. We read industrial and academic papers, textbooks, and historical and scholarly technical articles to get a better knowledge of the measuring techniques used for clothing. The following were the selection and review criteria: (a) The paper concentrates on clothing-specific measuring techniques, and (b) it offers enough information to ascertain the measurement procedure. The review findings are given as an overview of the measurement techniques arranged according to the methodology for gathering body data, together with an analysis of the techniques. We talk about the creation and use of the main measuring techniques in the garment development process within this framework. The world of clothes underwent a profound upheaval during the Industrial Revolution. The days of custom tailors making clothes to fit each individual were long gone. Factories that produced mass-produced readymade garments replaced them. This change brought with it a new challenge: how to guarantee that clothes fit a variety of bodies without the luxury of custom tailoring, even though it was more affordable and convenient? [4].

The Need for Uniformity

Before industrialization, choosing clothes was a private matter. Carefully measuring each customer, tailors made clothes that flowed and draped to suit each person's demands and tastes. But big production necessitated a different strategy. It was necessary for manufacturers to provide clothing in conventional sizes that would suit a variety of body shapes without requiring significant alterations. The earliest standardized size schemes were developed in response to this requirement for consistency.

Early Attempts at Standardization

The early attempts at standardizing size were far from ideal. A few important dimensions, including as waist, inseam, and chest circumference, were frequently the focus of early efforts. These dimensions were often based on pre-existing systems, such military sizing charts used for troop outfitting. In order to create baseline measurements, a few manufacturers also carried out brief anthropometric studies—the science of measuring the human body—on small sample sizes.

Standardized Measurements

Standardized measurements were required for purposes other than clothing. The uniform size of manufactured goods was essential to the success of the Industrial Revolution as a whole. Standardized wrench sizes were necessary for interchangeability, while furniture designers wanted chairs to have seats of a certain width for comfort. This focus on consistency made mass production easier and assembly procedures more efficient.

Challenges and Variations

Reaching genuinely global size standards remained a major obstacle in spite of these attempts. Confusion for customers resulted from different manufacturers using different systems. It's possible that a "medium" size from one manufacturer won't fit a "medium" from another. In addition, size systems that failed to account for the variety of human body shapes were frequently the consequence of the dependence on scant anthropometric data. These early systems of size frequently did not take into consideration the fact that people from different locations and ethnic backgrounds may have differing proportions.

The Birth of Modern Sizing Standards

Sizing standardization advanced significantly in the early-twentieth century. Researchers and statisticians started carrying out more thorough anthropometric analyses. This required collecting information from populations that were bigger and more varied in terms of age, ethnicity, and body type. This information served as the cornerstone for increasingly complex scaling schemes that sought to depict the human body more accurately.

Sizing Charts and Tables: A Visual Guide to Fit

A revolution in customer convenience was brought about with the introduction of sizing charts and tables. These charts gave standardized measurements, and the accompanying clothing sizes a visual representation. Sizing charts, which were usually shown in stores or on clothing tags, gave customers the ability to select garments according to their own body measurements, a level of self-selection not possible during the custom tailoring era [5].

World War II and Post-war Sizing: From the Battlefield to the Department Store

The significance of uniform sizing was further highlighted by the World War II, especially with regard to military gear. During this time, anthropometric data were collected extensively, providing important insights on the proportions and body sizes of soldiers. By using these data, sizing procedures were improved to guarantee that soldiers in the field would fit and perform properly.

Post-war Consumer Culture and the Ideal Body

Clothes became more than just a need during the post-war era of consumer culture boom. Ads and mass marketing frequently promoted "ideal" body forms, especially for females. Sizing conventions may have been impacted by this cultural movement toward a more limited definition of beauty. Some manufacturers may have changed sizing to accommodate these imagined norms, regardless of actual body proportions, in an effort to profit from these ideals. This might have caused a discrepancy in the way clothes fit a larger range of body types in reality from the sizes that were advertised.

Contemporary Sizing Systems: A Global Challenge, a Flawed System

Sizing systems now faces additional difficulties as a result of globalization and shifting populations. Populations differ greatly in terms of body types; therefore, a "one-size-fits-all" strategy is no longer effective. The popularity of internet shopping has also made choosing a size more difficult because it frequently prevents customers from trying on clothing before making a purchase.

Challenges and Criticisms: A System in Need of Repair

There is a lot of criticism directed toward the current size methods since they are not inclusive, inconsistent, or accurate. "Vanity sizing"—the practice of identifying clothing with a smaller size tag than the actual measurements—is a common problem that irritates and confuses customers. Moreover, size systems frequently ignore the varied needs of the world's population in favor of accommodating a narrow range of body shapes. Finding securely fitting clothing can be difficult for those with non-normative body shapes, lower or bigger statures, and other body types [5].

Future Trends in Sizing Systems

Innovation is moving toward a future where inclusion and personalization are given priority because of the shortcomings of the existing sizing methods. Here's a look at what could come next:

Technology and Personalized Sizing: A Tailor in Your Pocket

Technological developments have significant opportunities to completely transform the sizing industry. The use of 3D body scanning technologies could lead to a paradigm change. Imagine using your smartphone at home or at a store, and having a 3D scan take a picture of your exact measurements and body shape. Afterward, a customized digital model might be created from this scan, enabling you to virtually try on clothing and make sure it fits well before making a purchase. Online shops can also incorporate 3D scanning technology into their platforms so that clients may upload scans and obtain recommendations for garments tailored to their unique body types [6].

Inclusive and Adaptive Sizing: Catering to a Diverse World

The acceptance of diversity is the key to the future of scaling systems. This entails creating standards that take into account the wide range of body shapes and races seen throughout the world. Uniformity does not always equate to standardization. It may entail expanding the selection of sizes available within pre-existing systems or perhaps constructing whole new sizing frameworks that more accurately capture the diversity of human proportions.

Adaptive Sizing Features: Clothes that Grow with You

Another interesting development is adaptive scaling features. Imagine wearing clothes that can be modified to fit a greater variety of body shapes and abilities, such as waist adjustments, expanding closures, or convertible necklines. People who fluctuate in weight, have physical restrictions, or need clothing that meets certain medical needs may find this very helpful.

The Rise of Sustainable Practices: Sizing Systems with a Conscience

Because of the present "fast fashion" model's inconsistent sizing and poor fit, textile waste is frequently increased. Future sizing schemes that guarantee longer-lasting and better-fitting clothing can help to encourage sustainable practices. Furthermore, the emergence of on-demand manufacturing, in which garments are made in response to individual orders and customized sizing information, may lessen textile waste and overproduction.

Challenges and Considerations

Even with the promise of these developments, difficulties still exist. To make 3D scanning technology widely available to a larger population, it could be necessary to expand infrastructure and lower costs. To win over customers, issues with data security and privacy pertaining to the gathering and archiving of body scan data will also need to be resolved.

Conclusion

The development of sizing systems is an intriguing tale of human inventiveness and the never-ending search for the ideal fit. One thing will always exist, regardless of how technology advances shape the future or how custom tailoring of the past shapes the present: the need for apparel that fits well, looks nice, and expresses our unique styles. To make clothes more effective, customized, and comfortable for everyone, we must embrace technology, diversity, and sustainable methods going forward.

References

1. L. Griffin, C. Compton, L.E. Dunne, An analysis of the variability of anatomical body references within ready-to-wear garment sizes, in *Proceedings of the 2016 ACM International Symposium on Wearable Computers* (ACM, New York, NY, USA), Sept. 2016, pp. 84–91. https://doi.org/10.1145/2971763.2971800

2. S. Gill, A review of research and innovation in garment sizing, prototyping and fitting. Text. Prog. **47**(1), 1–85 (2015). https://doi.org/10.1080/00405167.2015.1023512
3. R. Florida, The new industrial revolution. Futures **23**(6), 559–576 (1991). https://doi.org/10.1016/0016-3287(91)90079-H
4. D.E. Veitch, M. Henneberg, Sizing for the clothing industry using principal component analysis-an australian example, 2005. [Online]. Available: www.astm.org
5. S. Xia, C. Istook, A method to create body sizing systems. Cloth. Text. Res. J. **35**(4), 235–248 (2017). https://doi.org/10.1177/0887302X17713298
6. S.P. Ashdown, L. Dunne, A study of automated custom fit: readiness of the technology for the apparel industry

Chapter 3
Development of Sizing System

Jawad Naeem

Abstract The creation of sizing system is very critical in making sure that garments and other items fit in proper and comfortable manner for the wearers. The procedure inculcates determination of standard measurements for several body shapes and types along with development of size charts which accurately depict these measurements. The sizing system must consider easiness in movement and construction of garment to make sure that fit of the garment is good. This chapter inculcates conductance of previous studies on measurement of the body and portions on several range of consumers along with important terms in sizing system. In addition, it also deals with the prominence of sizing system and reviewing sizing system which had been developed for previous studies. Furthermore, this chapter exhibits the creation of sizing system along with statistical methodology and data mining practices. In consequence, the creation of a sizing system requires cautious contemplation of several parameters to make sure that fitting demand of consumers is fulfilled.

Keywords Size range · Size scale · Drop · Size designation · Control dimension

Introduction

When the mass production of garments started, sizing of human body was utilized for segregating standardized body and clothing measurements into several categories. The main aim was to improve productivity and selling of clothing [1]. Dickerson [2] mentioned the standardization of clothing is a pivotal issue for ready to wear clothing, as it leads to formation of various sets of human body dimensions with same designation of size. From the historical point of view, the garment manufacturing industry established their own sizing methodologies for ready to wear clothing. Due to this reason, there was too much misunderstanding along with variance of sizes in clothing [3]. At present, industry is doing its level best to fulfill clothing fit demand

J. Naeem (✉)
Department of Textile Engineering Technology, School of Engineering and Technology, National Textile University Faisalabad, Faisalabad, Pakistan
e-mail: jawad@ntu.edu.pk

M. Q. Khan et al. (eds.), *Garment Sizing and Pattern Making*, SDGs and Textiles,
https://doi.org/10.1007/978-981-97-7683-2_3

of customers due to preferences of customers toward acute clothing fit with respect to human body [4, 5].

Before the industrial revolution, garments were stitched with hands according to the measurements of customers along with his preferences of color, shape, and style. It was the common practice in almost every culture and was affected by the social, economic, and traditional customs all over the world. In Europe, the production of garments took place in home workshops. The scale of shops was small. The clothing was stitched with hands as per accurate measurements of individual customers. Due to this reason, there was a problem related to sizes and fittings of garments [6].

These domestic workshops were dependent on skills of tailors, stitching workers and seamstresses who were responsible for making these garments with different styles, designs, and qualities. Several months were consumed for making these garments and were worn for lifetime. These clothes were transferred to next generations [7].

The Industrial Revolution offered swift production systems with the help of sewing machines and its advanced mechanical tools. On the other hand, it had some negative influence on economic prospects of tailors, skilled workers, and clothing business of Britain [8]. Almost similar situation was faced by craftsperson of India. Many royal workshops were utilized as institutions for providing knowledge to the young generation. This paved the way for lot of experimentation of different technologies for textiles and apparel [9].

Initially, the method of sizing for production of alike garment in batches was done through the gathering of anthropometric data, i.e., in case of men's military uniform during the Crimean war and American revolutionary wars [10–12]. This methodology laid the basis of sizing system for determining whether the garment is loose or tightly fit to the body of wearer [13]. The growth of newly industrial estates in America and Europe enhanced the need of formal work wear for urban society who were working in several disciplines of production, administration, and finance [10]. As a result, a new approach was established for the retail sector business. Initially, the larger shops, warehouses, and department stores were established to cater the clothing needs of men. The main reason behind this was the simplicity in the style of men's clothing as compared to women's clothing. Moreover, women's body is more complicated than men's body as men's clothing required the dimensions of shoulder and chest for making of shirts and coats [10].

Development of Sizing System

A sizing system is a table having numbers which depicts values of each dimension of the body utilized for classification of human bodies in the population for each size group in system [14]. With the enhancement of demand of women's wear after the end of World War I, there was demand of regular sizing system and standardization of different groups and subgroups of target population [10]. The main aim of sizing system for women's clothing was to deliver fitted fashion garments to a huge group of population having similar proportions of body under the same dimensional standards [15–17]. The sizing system inculcates two types of methods. In one method, development of size system was based on anthropometric data of population groups from several countries and in other methodology size standards were established by various clothing brands for their target customer.

Ashdown mentioned four types of sizing system which are as under:

i. One size fit

It inculcates various design features and characteristics of materials like several knits for sportswear, t shirts etc.

ii. Standard sizing system

This system mentioned various range of sizes from the globe as small, medium, large, or alpha sizing system with a smaller number of size range and a numerical sizing, i.e., 6–8–10–12–14–16, etc. covering up large size range.

iii. Mass customization sizing system

In this system, the customer/wearing person was presented with various variables along with dimensions of clothing like short, regular, long in small, medium, and large sizes. As a result of that additional options were available for consumers.

iv. Custom sizing system

In this system, personal size was made for every wearer [18].

In order to develop sizing systems, there is possibility of series of decisions [14].

i. Controlling dimensions, i.e., depending on garment, which part of human body and how many human body measurements are employed for classification of population.
ii. Establishing the size range, i.e., what proportion of range of values must be included for every measurement of human body from the population.
iii. Formation of size, inter size interval or size step, i.e., how the range of values of every dimension must be segregated into groups and size groups.
iv. Creation of size roll, i.e., how many sizes must be generated and number of garments that must be manufactured for each style.
v. Decision of secondary measurements of human body that are significant for preparation of garment.
vi. Designation of sizes for labeling of garment for identification of consumer to avoid any confusion [14].

Foundations of Current International Sizing System

Development of the sizing system is based on feedback from the wearer, evaluation of sales and customer feedback reports. The adjustment of the sizing system was done by trial-and-error methodology. Sometimes modifications in the measurements of clothing for peculiar sizes were done without alteration in size designation. Due to this reason, there was variation in sizes of clothing for different companies which indicated that garments with same measurement might have different size numbers. Size designations might be referred to as code numbers which do not represent human body or clothing dimensions. This resulted in unsatisfactory customers and unpleasant feelings toward their body [19].

Standardization institutes must take the initiative to develop sizing system for population of country using data of anthropometric studies to avoid confusion between customer and industry. These systems must be able to identify sizes and shapes of much of the population. Moreover, each size group must be able to establish tabulation of human body dimensions. The label referring to size designation must highlight these body dimensions. In this way, consumers can swiftly uncover garments according to his own fitting. Those markets which are not covered by the standardized sizing systems, clothing brands must design their own sizing systems to satisfy these customers [14].

It is imperative to focus on differentiation between clothing sizing and human body sizing, i.e., addition of amount of fabric above or beyond several landmarks of body known as garment ease. It indicates the fit of the garment and is decided by the designer. The garment sizing system organizes measurements of clothing which will fit the body sizing system after the setting of ease allowance by fashion designer. The deviation in ease allowance from one garment to another garment is the possible explanation of clothing assigned with same size have varying dimensions [20–22].

In 1968, the Swedish member of international organization for standardization suggested for Technical Committee for development of international sizing system for clothing [23]. Despite participation of 30 countries, the initial findings from several countries showed that establishing a global sizing system was not viable due to significant variations in population. Thus, development of a single sizing system for accommodating the population of whole world would not be affective because insignificant portion of sizes would be employed specific country owing to its ethnicity. As a result, the working of ISO/TC133 directed their efforts on standardizing sizing system instead of developing international standard sizes [23].

Technical reports issued by ISO/TC 133 mentioned selected key measurements, values, intersize intervals for every measurement and establish ways for development of sizing system on basis of anthropometric data for specific population. The report mentioned instructions on creation of sizing systems from anthropometric data collected according to ISO 8559:1989 standard [23].

Countries like Japan, England, South Korea, and Hungry [24] had improved their size designation systems as standards of ISO. For example, JIS L 4005:2001 sizing standard was developed in Japan in accordance with ISO/TR 10652:1991 by Japanese standard association in 2001. The European committee for standardization established series of standards EN 13402 for defining body dimensions and procedures, measurements establishing the sizes of several types of garments and acknowledgment of garment sizes. These standards practiced the pictogram established by ISO and for designation of sizes, code was extracted from numerical values of measurements [14].

Important Terminologies in Sizing System

Key dimensions

The composition of sizing systems is built on segregation of population into groups with almost alike body dimensions. These dimensions of the body are employed to categorize population into groups called control key dimensions or control dimensions [14].

Primary control dimensions

The primary control dimensions break up population into big size groups aside measurement of the body for peculiar type of clothing. For example, if length/height of the body is chosen as primary control dimension, people will be divided into height groups [14].

Secondary control dimensions

Afterward, each size group is divided into subgroups rendering to a secondary control dimension which is second important control dimension for the development of particular type of clothing. If the secondary control dimension is hip girth, then the people in each height group is categorized into subgroups accompanied by measurement of hip [14].

Tertiary control dimensions

The subgroups can be expanded into a category called tertiary control dimension, i.e., distinguishing the body into peculiar proportions called body shape or body type. Thus, the additional subdivision of group describes more detail of body in terms of shapes of size group [14].

Size range

Every control dimension is associated with typical body measurement, whose values in the population alter between some minimum and maximum values. These set of values enclosed in sizing chart are called size range alongside that control dimension [14].

Accommodation rate

If the size range is lesser as compared to population range, then some persons will not be included in sizing system. Due to this reason, they will not be able to acquire ready to wear garments according to their dimensions. The section of population given for the sizing system is known as accommodation rate of sizing system. The accommodation rate is mostly between 65 and 85% and is dependent on type of clothing. The accommodation rate can be assessed by using several control dimensions and correlation coefficients among them [25].

Size scale

The range of every control dimension is further segregated by developing a set of sizes called size scale. The size scale is dependent on enhancement between adjoining sizes. This enhancement, i.e., size interval, size step, or size grade, may have constant or variable value [14].

Drop

The relation between two body dimensions based on identification of body type is known as drop of sizing system. This association can inculcate either secondary or controlled dimensions and might be utilized to distinguish common body shape groups in population, i.e., several body shapes will relate to various values of system's drop. In most cases, the drop is expressed by differentiation between hip and bust girths for women. However, for men the drop value is the difference between chest and waist girths [14].

Size designation

The coding system for the sizes that permit the consumer to distinguish the body dimensions for which the garment was developed is known as labeling or size designation. The methodology of size designation along with size of every group will deliver certain accommodation rate of population establishing the sizing system, size roll, or traffic system for particular type of clothing [14].

Traditional Methodologies of Sizing System

Starts with the gathering of anthropometric data about the population being studied. As it is not possible to evaluate each single person, a cautious approach is needed for selection of sample which is statistical representation of population trailed by adopting appropriate statistical measures. The methodologies of taking typical measurements are controlled with standards like ISO 8559:1989, ASTM D5219, CS 215-58 or PS 42-70 or EN 13402-1 [14].

The development of sizing system goes through these following major steps:

i. Choosing control and secondary dimensions with respect to garment type and definition of body types in the population.
ii. Selection of inter size intervals and description of scales for control measurements.
iii. Formation of the optimum number of sizes.
iv. Computation of the values of secondary dimensions [14].

Choosing Control Dimensions

The quantity of sizes is linked to amount of control dimensions and scales of control dimensions. As differences in body shapes in population are huge, the number of controlled dimensions required for illustration of body specifics enhances. As a result, there was huge increase in the number of sizes required to adjust population. The number of sizes can be reduced only if the portion of body covered by the clothing is employed for classification. The categorizing of the shapes will be simpler if clothing is expected to cover only the upper body (shirts and jackets) or the lower body parts (skirts and pants) or just for head like hats. In this case, a small number of sizes are required. In consequence, size charts are developed by only considering appropriate types of clothing. Explanation of shape of body portion wrapped by only peculiar type of garment means that only body dimensions inculcated in development of garment must be contemplated for control or secondary dimensions. ISO standard 3637:1977 mentioned for whole or top body garment to employed three controlled dimensions, i.e., bust girth, hip girth, and height lest the garment is swim wear or knitwear, in that situation only two body dimensions are suggested, i.e., bust and hip girth for swimwear and bust girth and height girth for knitwear [26].

Koblyakova [25] in 1980 mentioned that control dimensions should have highest absolute value to represent basic shape of the body. This proposition can be comprehended only by keeping in mind a group of objects from a large distance. The first attribute with which shapes of the object might be identified will be the larger dimension. It will permit detection of objects in short and tall form. Afterward, the next largest dimensions will play its part to distinguish the objects and their shapes in wide and narrow form. Koblyakova also mentioned that chosen controlled dimensions must be needed to lay in different planes because measurements might be linked to other measurements in parallel or same plane. However, previous studies mentioned that horizontal and vertical measurements do not correlate well with each other [27]. In addition, the control dimensions must be chosen in such a way that every secondary dimension must be correlated properly to minimum one of the key measurements [3, 23, 25, 28, 29]. This correlation can be employed to establish a complete set of body measurements for every size group. The complete set of measurements shall deliver data for development of patterns of clothing for every size. In the end, the approach will be more practical if selected control measurements were easy to choose [28, 29]. For example, length of crotch is not good approach for control measurements as it is complicated and inaccurate to measure on wearable person particularly in case of women wearing skirts. In addition to that, the selected measurements should be more acquainted with customers, if sizing system employ coding system built on control measurements. Yu [30] in 2004 mentioned that in previous century, statistical analysis of anthropometric data was carried out in a lot of studies. The outcomes of these previous studies might differ from one population to another and from year to year. However, the major outcomes about the proportions of the body did not vary too much. These stipulatory controlled dimensions can be learned in the documentation form published by international organization for standardization on sizes of clothing and size designation [14].

Very first scientific study of body measurements employed on development of women's clothing was performed by O'Brien and Shelton in 1941 [28]. In this study, they gathered fifty-nine human body measurements from ten thousand and forty-two females in America. Basic statistics like medians, modes and standard deviations were used for data analysis followed by bivariate distributions along with frequency distributions. To create a relationship between variables, a correlation coefficient was calculated. A comparison of correlation coefficients between pairs may induce to those variables which associate with many variables simultaneously. The outcomes were validated with further statistical analyses like principal component analysis and partial correlations [14].

Selection of Inter Size Intervals

After the control and secondary dimensions of sizing systems and types of body confronted in population were evaluated, the further step was the formation of size groups. Selection of inter size interval along with every key dimension and separating the range of measurement by this value also established the number of sizes. Therefore, the value of intersize interval along with anticipated accommodation range specifies the number of sizes. At the first stage, range of coverage must be determined, or accommodation rate provided by sizing system. After examination of spread of values in population, calculation of percentile values can be done. For example, if 45th percentile for men's size was 1.43 m, then forty five percent of male in population are 1.43 m or less tall. Percentiles can be employed to evaluate the percentage of population which will be covered by the minimum and maximum values of given accommodated. If eighty percent of population is covered, a design may be made for size range to accommodate values between 5 and 85th percentiles or between 10 and 19th percentiles. In that scenario, 20 percent of population will not be included in sizing system [14].

In a study conducted by Ashdown and Delong in [31], a series of perception fit test were performed, and it was inferred that threshold level (tolerance level) for trousers can be 0.5 cm for waist and 1.5 cm for hip and crotch length. However, the size of study was insignificant to develop statistically for validation of confirmation of values. Perception fit tests are complicated to perform, they are usually not carried out for the establishment of sizing system. The steps of sizing system were established on the basis of easiness, convenience and customs [32]. It is sometimes improbable to compare two sizing systems using same alike control dimensions and units because of differentiation in steps and scale of sizing system. Due to this reason, ISO issued a set of chosen size intervals, size scales, and matching points along scales [33]. If chosen numbers are utilized, several sizing systems can be conveniently associated, and their equivalence can be developed.

Establishing the Number of Sizes

Calculations of size interval and range of every dimension to be accommodated evaluates complete set of sizes which will cover entire population. The set of sizes was decided by combination of all probable values which control dimensions can assume. From all these set of sizes, only those sizes were selected for production, which illustrated sections of population above certain percentage. In this way, the number of sizes in the sizing system can be characterized. If the size step is defined, there are two ways to resolve the problem: one way is to find optimal number of sizes under accommodation rate, and second way is to search for the accommodation rate according to number of sizes. To resolve this problem, proper analysis of anthropometric data was required and for acquiring the joint frequency distributions for the size groups mentioned by selected control dimensions and their comparable size steps [14].

Determination of Secondary Dimensions

The secondary dimensions illustrated a body in comprehensive details required to develop some clothing to fit a human body. By analyzing the human body data of population, it was noted selection of control dimensions are significantly associated with the rest of the measurements, giving them the predictability attributes. Illustration of dimension as functionality of control dimensions permit projection of its values relatively well for any peculiar combination of values for control dimensions. The impact of every important dimension into the sets demonstrating a particular measurement is evaluated through multiple regression analysis by utilizing data of population. The outcomes of multiple regression process are combination of coefficients related to every measurement [14].

If there is linear regression on key dimensions measurement, there will be only single coefficient in combination of regression coefficient for every important dimensions. The measurement will be proportionate to the important dimensions, and subsequent sizing system will be called proportional sizing system [14].

Some Other Ways to Construct Sizing Systems

In a study conducted by Maria et al. [34], a new methodology was designed for development of sizing system for mass customization of Greek men. This new methodology was obtained on the basis of statistical analysis from anthropometric data which was assisted by model of iterative mass customization. As a result, there was the introduction of satisfaction performance indices. The methodology derived was effectively implemented to anthropometric data comprising of twelve thousand and eight hundred and ten Greek men [34].

In statistical analysis, minimum, maximum, mean value, and standard deviation were used for discrete combination of body measurements. Least square linear regression methodology was used, and correlation among several body dimensions was find out. Target population was segregated in different body types as per body measurements of every subject. Height dimension and drop value was used for segregation. Drop value was = Chest girth − Waist girth [35, 36]. The drop value was employed to segregate the population by measuring the specific association among key dimensions [35, 37]. Secondary size was evaluated for every primary size with respect to type of garment and limitations. In the same way, for every primary dimension there is also a secondary dimension and number of acquired secondary sizes n was evaluated through the following equation:

$$n = \frac{(\text{Mean} + 2\text{SD}) - (\text{Mean} - 2\text{SD})}{\delta} = \frac{4\text{SD}}{\delta} \tag{3.1}$$

Every mass customization model describes a sizing system which initiated with medium shape body and continues to cover nearby body shapes in corresponding to this model. Each model is dependent on the anticipated mass customization degree which is improved in iterative manner in sequence of steps. For acquiring every mass customization model, sizing system was established according to following arrangement [34]:

i. Mass production

Developing medium body sizes, i.e., primary sizes, which are utilized for traditional mass production Prêt-à-porter clothing.

ii. Model one for mass customization

Developing two secondary sizes for each primary size in accordance with 1st secondary dimension.

iii. Model two for mass customization

Developing eight secondary sizes for every primary sizes in accordance with 1st and 2nd secondary dimensions. The validity of this model is suitable for garments having two secondary dimensions.

iv. Model three for mass customization

This step was divided into two cases: Case A and Case B.

In case A, there is only one secondary dimension, and four secondary sizes were developed for each primary size in accordance with the secondary dimension.

Whereas in case B, there were at least two secondary dimensions. Twenty-five sizes were developed together with two secondary dimensions for every primary size. These sizes were generated for two body classes which were before and after medium size body shape as mentioned in Fig. 3.1

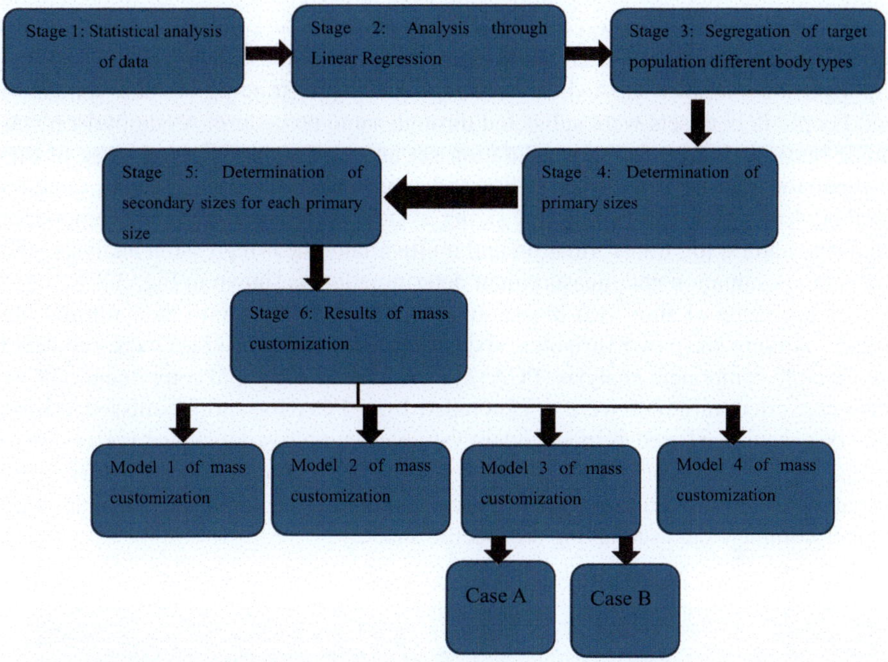

Fig. 3.1 Flow diagram of suggested mass customization methodology [34]

v. Model four for mass customization

Developing all probable sizes for each primary and secondary dimension. That model was related to a sizing system with maximum agreement of population and a greater number of several clothing sizes. The outcome of experimentation showed that developed garments fit to 99.9% of the target population [34]. The suggested method for developing a sizing system together with computer-aided and information technologies might assist in the development of garments, and customization for fit in a swift manner. These customized clothing can be induced in unit production lines as added size and manufactured like every other garment having the same style. This indicated that prosperous organizations having several styles of garments would be capable of implementing the suggested mass customization strategy with moderately minor endeavors. Prospective enhancement in cost manufacturing by cutting limited garments at one time instead of hundreds might be compensated with increment of sales and faithfulness of customers [34]. The suggested methodology for mass customization might be applied by utilizing either one manual measurement or automatic scanning system. However, traditional methods have some minor errors and noise. On the other hand, these discrepancies were not presumed to occur in automatic scanning data which is extremely precise and filtered for noise elimination [34].

In a study conducted by Xia [38], a sizing system was created for several body shapes based on USA size anthropometric data. Six thousand three hundred and eight persons in sized USA data were segregated into training set and validation set. People in both sets were subjected through same procedure. A comparison was made between outcomes from validation set and training set. The purpose of this comparison was to test the durability and repeatability of sizing system creation methodology. The development of sizing system inculcated principal component analysis, natural log transformation and multivariate linear regression analysis, and size range evaluation and measurement determination as shown in Fig. 3.2.

An aggregate of sixty-two measurements were converted into their normal log values. Among sixty-two variables, sixty variables with 2 variables were subjected to principle component analysis (PCA) and evaluated 2 principal components (PCs). Two converted variables were not included from PCA for various missed values. The PCs were employed as independent variables in multivariate linear regression to predict all sixty-four measurements. Multivariate linear regression was performed on rectangular bottom hourglass body shapes, spoon body shapes, and rectangle body shapes complete data set having no specific shape.

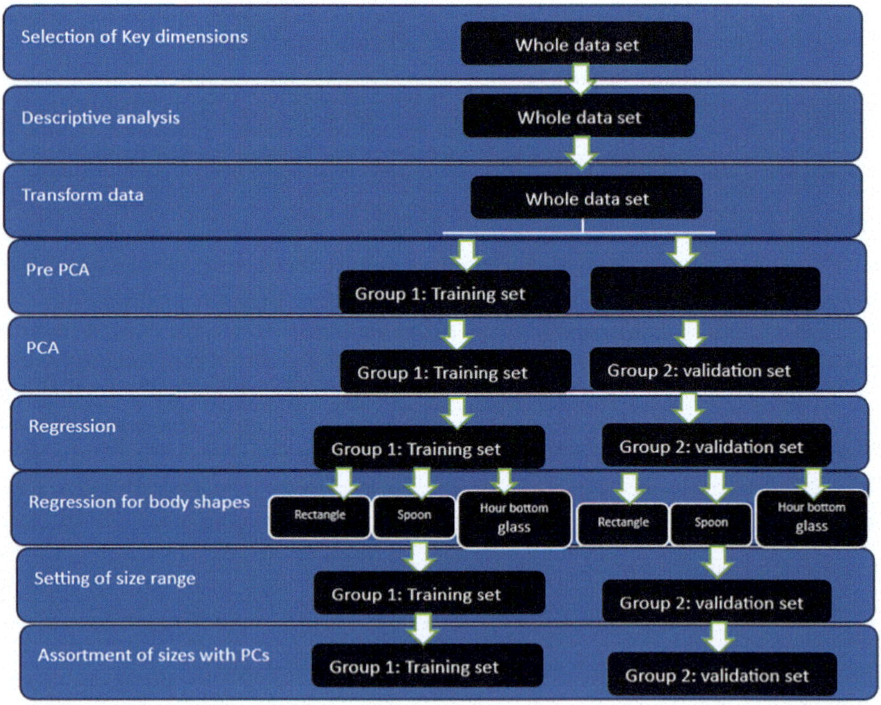

Fig. 3.2 Principal component analysis, natural log transformation and multivariate linear regression analysis, and size range evaluation and measurement

Sizes were evaluated by grouping principal components in range decided by standard deviation and mean values. Principal component 1 was segregated into fourteen groups uniformly within a range adjusted to [mean − 2.3 SD, mean + 2SD]. The principal component 2 was segregated into three groups uniformly within the range set to [mean − 2SD, mean + 2SD]. The midpoint principal component values for all sizes were implied to formularies created in multivariate linear regression investigation to determine measurement values. A comparison of developed sizing system was then made with ASTM D5585-11 [39]. Investigation of the outcomes determined that there was repeatability and reliability in the methodology for development of sizing system. Improved fit was acquired in developed sizing system for persons from validation sets than ASTM D5585-11. As a result, it was imperative to inculcate body formation in sizing systems [38].

Hammad et al. [40] developed a new methodology to reduce number of sizes by testing on data collected from France. The methodology might also be utilized on some other anthropometric data sets and assisted in developing sizing systems with reduced number of sizes suitable for garment industry in terms of cost. The methodology inculcates a procedure of choosing important measurements, size interval established on already issued standards. After the development of sizing system, the algorithm was employed for determination of neighboring matrix. A very accurate deleted arrangement of sizes was executed. In addition, a set of several size intervals was implemented, and a comparison was made to find which set worked in best way. Furthermore, the research was used to investigate best approach to employ sizing system on the group resulting from clusters but not the entire population [40].

The aim of the research conducted by Ola et al. [41] was to build statistical model for important body dimensions, i.e., hip, waist, and bust for making size chart of ready to wear garments for pear shaped South African women. Correlation methodology was utilized for body measurements from suggested chosen women having ages from twenty-five to fifty-five years. After segregation of height measurements into 3 groups, the means of body measurements for medium height group were employed to build size charts from size sixteen to size twenty-four by employing principal component analysis (PCA) and least square regression methodology. Outcomes indicated that bust, waist, and hip values significantly participated in body type sizing. The bust and hip were considerably correlated at $R^2 = 0.996$ (99.6%) and model anticipated the true Value of hip at $R^2 = 0.993$ (99.3%). Bust measurement was positive correlated to waist at 93.8% and the model anticipated the true value of the waist at $R^2 = 0.880$ (88%). Bust measurement was a considerable factor in forecasting the hip and waist dimensions. It was also noted that the outcomes of simulated values for waist and hip measurements at several sizes dependent on bust measurements proposed that values acquired correspondingly were quite like the measurements in customized size chart obtained during study of Ola et al. [41].

Majidi et al. [42] in his research tried to create sizing system for high school students of Iran. Anthropometric data of three hundred and eighty male students was acquired. The ages varied from fifteen to seventeen years. Cluster-based data mining methodology was utilized. The aim was to inculcate a higher percentage of population by employing few sizes and offering the industrialist with point of reference to market their products in accordance with type of body and dissemination of sizes. There is dire need to update the anthropometric data in a continuous manner; therefore the size charts must be rebuilt in continuous manner through the application of suggested data mining cycle [42].

Destaw [43] conducted research to develop sizing system for the girls of Ethiopian Amahara from age five to ten years for manufacturing comfortable and fitted clothes. Correlation coefficients were employed to measure pivotal dimensions to fulfill the principal of significant associations with several dimensions along with the portion of body. Dimensions of height, chest, hip, side waist to knee and outside leg were chosen as standardized norm for sizing, verifying centilong system for specimens under investigation. It was inferred from the study that a set of percentiles was pertinent to create the size chart for clothing according to the fits [43].

Zakaria and Gupta [15] mentioned a complete method for creation of sizing system as depicted in Fig. 3.3. The method was divided into three stages. Stage one is analysis of anthropometric data, in which preparation of field work, anthropometric planning and survey along with analysis of anthropometric data took place. Stage two is related to sizing analysis including multivariate analysis, principal component analysis, cluster analysis and decision tree analysis. Stage three is related to the creation of a sizing system. This stage included size system development, validation of sizing system and designation of sizing system.

Modification and Adjustment of Sizing System

It is understood that all factors of sizing system are dependent on the data which served as basis of sizing system and subsequently on the group of persons who were measured. If a sizing system is utilized by the population apart from that measured in original manner, system will requisite modification. Body measurements of the new population will be collected, and comparison will be made with the original set of data. If the differentiation is significant in a statistical manner, it must be induced into novel body sizing system. Modification in sizing system might involve moving the size scales in certain arrangement to inculcate persons with measurements focus on both ends initial sizing system, addition or removal of size groups determined by further joint distributions, modifications in size intervals to squeeze the system or opening it up, presenting new body types and removing others etc. The problem in this situation utilizes the similar practices like establishing a whole new sizing system [14].

The suppositions that clothing sizes imitate body sizes had focused the research on illustration of wearers body and developing body sizing systems, which are shifted toward garments for development of sizing system. On the other hand, garment fit is dependent on the addition of ease allowances above and beyond measurements of wearer's body for comfort and style. The quantity of ease allowance was not equivalent for all sizes. For ideal scenario, to improve the sizing system, a clothing pattern can be developed for each size and fit of garment is tested and modified

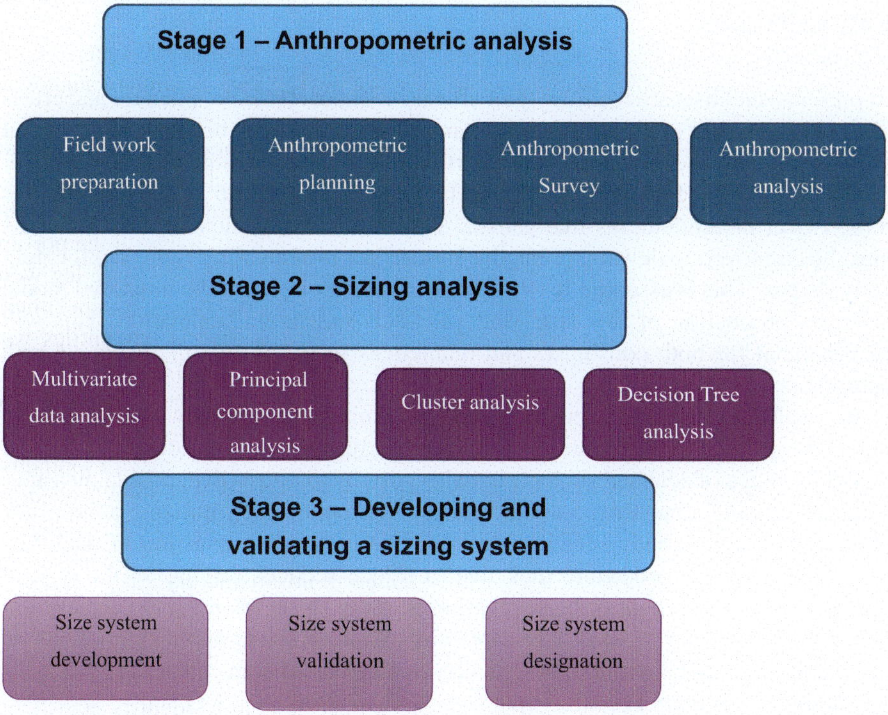

Fig. 3.3 Flow diagram of methodology of creation of sizing system [15]

on wearer's body which represented that body size group. After the adjustment of garment patterns for all sizes, the garment sizing system having suitable experimentally driven ease allowance will be obvious. In this way, garment sizing system will not essentially reproduce the inter size intervals of body sizing systems which it fits [44].

Modifications in sizing systems for population must yield improved fit for garments for that typical population. However, with some other sizing system, the difficulties of size designation and identification must be tackled in a very effective manner. A transparent approach for communing the dimensions of the wearer for whom the clothing of a particular size had been developed is essential and standardization of size designation as suggested by ISO is a very anticipated resolution to solve the problem.

Future Trends

At present, scientists are focusing their efforts on improvement of efficient and effective clothing sizing systems. The important efficiencies occur between size roll and accommodation rate. Advancement in artificial intelligence must pave the way for producing improved sizing system with better accommodation rate and lower size rolls. Novel methodologies like genetic algorithms and artificial neural networks must be employed to develop predictive models for investigating right size for right body shapes. This is essential for the garment industry because the improved model indicates production of few sizes and yet holds maximum population. This will be beneficial for both consumers and retailers as model fulfills the needs of both consumers and retailers [15].

It is believed that advancement in body scanning technology would have huge influence on creation of sizing systems. These body scanners will not only bring real time 3-D image of scanned subjects but also permits to obtain dependable measurements across and along the scanned surface of human body by utilizing appropriate software. Complete body scanner is established for swift extrication of data, i.e., the whole scanning procedure took five to fifteen seconds relying on the type of scanner and development of body measurements from the scan can be achieved in a time less than sixty seconds. The capability of taking whole duplication of human body shape in small amount of time is very attractive and convenience approach for anthropometric survey [14, 45, 46].

References

1. A. Beazley, Size and fit: procedures in undertaking a survey of body measurements. J. Fashion Mark. Manage. **2**(1), 55–85 (1997)
2. K.G. Dickerson, *Inside the Fashion Business*, 7th ed. (Prentice Hall, New Jersey, 2000)
3. J.M. Winks, *Clothing Sizes: International Standardization* (Redwood Book Textile Institute, Manchester, 1997)
4. R. Otieno, *New clothing sizes charts for 3–6 years old female nursery school children in the Nairobi Province of Kenya: implications for marketing strategy* (Metropolitan University, Manchester, 1998)
5. S. Adu-Boakye, J. Power, T. Wallace, Z. Chen, H. Faculty, Development of a sizing system for Ghanaian women for the production of ready-to-wear clothing
6. Connor-Crabb, History of sizing systems and ready to wear garments 2007. PhD thesis, Aldrich, 2007
7. Connor-Crabb, A. Claire, Fashion design for longevity: design strategies and their implementation in practice. PhD Thesis report, University of Brighton, 330, Brighton, 2017
8. D. Monet, Ready-to-wear: a short history of the garment industry (2022)
9. R. Kumar, Costumes and textiles of Royal India. *Antique Collectors' Club* (2006)
10. W. Aldrich, History of sizing systems and ready-to-wear garments, in *Sizing in Clothing*, ed. by S.P. Ashdown (Woodhead Publishing, Cambridge, 2007)
11. A. Godley, The development of the clothing industry: technology and fashion textile history. **26**(1), 10 (1997)
12. P. Reena, W. Yu, Garment sizing and fit, in *Garment Manufacturing Technology*, ed. by R. Padhye, R. Nayak (Woodhead Publishing, Cambridge, 2015), pp. 187–204

13. W.L. Todd, Sizing for the military, in *Sizing in Clothing*. ed. by S.P. Ashdown (Woodhead Publication, Cambridge, 2007), pp.277–308
14. A. Petrova, Creating sizing systems, in *Sizing in Clothing*. ed. by S.P. Ashdown (Woodhead Publishing, Cambridge, 2007), p.30
15. N. Zakaria, D. Gupta, Apparel sizing: existing sizing systems and the development of new sizing systems, in *Anthropometry, Apparel Sizing and Design* (Elsevier Ltd., 2014), pp. 3–33. https://doi.org/10.1533/9780857096890.1.3
16. D.H. Branson, J. Nam, Materials and sizing, in *Sizing in Clothing*, 12th ed., ed. by A. Susan (Woodhead Publishing, Cambridge, 2007)
17. M.A. Mastamet, M. Klerk Helena, S. Jaqui, P. Ashdown Susan, Consumer' knowledge on sizing and fit issues: a solution to successful apparel selection in developing countries. Int. J. Cons. Stud. **10** (2008)
18. S. Ashdown, *Sizing in Clothing* (Woodhead Publishing, Cambridge, 2007)
19. K.L. LaBat, M.R. DeLong, Body cathexis and satisfaction with fit of apparel. Cloth. Text. Res. J. **8**(2), 43–48 (1990)
20. R. Feitelberg, Industry debate: are sizing standards needed? *Women's Wear Daily* (1998)
21. K. DesMarteau, Pre-production and CAD: Let the fit revolution begin. Bobbin **42**(2), 42–56 (2000)
22. J.E. Workman, E.S. Lentz, Measurement specifications for manufacturers' prototype bodies. Cloth. Text. Res. J. **18**(4), 251–259 (2000)
23. J.M. Winks, Clothing sizes, in *International Standardization* (1997)
24. J. Chun-Yoon, C.R. Jasper, Garment sizing systems—an international comparison. Int. J. Cloth. Sci. Technol. **5**(5), 28–37 (1993)
25. E.B. Koblyakova, *Osnovi Konstruirovaniya Odejdi* (Legkaya Industriya, Moscow, 1980)
26. International Organization for Standardization. ISO 3637:1977 Size Designation of Clothes—Women's and Girl's Outerwear Garments, International Organization for Standardization (1977)
27. E.B. Koblyakova, Razmernaya Tipologijya Naseleniya Stran—Chlenov SEV, Moscow (1974)
28. R. O'Brien, W. Shelton, Women's measurements for garment and pattern construction, in *Miscellaneous Publication 454* (US Department of Agriculture, Washington, DC., 1941)
29. J. Fan, W. Yu, L. Hunter, *Clothing Appearance and Fit: Science and Technology* (Woodhead Publishing, Cambridge., 2004)
30. W. Yu, Human anthropometrics and sizing systems, in *Clothing Appearance and Fit: Science and Technology*, ed. by W. Yu, L.H.J. Fan (Woodhead Publishing, Cambridge, 2004), pp. 169–195
31. S. Ashdown, M. DeLong, Perception testing of apparel ease variation. Appl. Ergon. **26**(1), 47–54 (1995)
32. Anon, *Plan Ahead*. Metrication and the U.S. Apparel Industry, Technical Advisory Committee, American Apparel Manufacturers Association, Arlington, Virginia (1975)
33. International Organization for Standardization. Standard sizing systems for clothes. Technical Report ISO/TR 10652:1991, Geneva (1991)
34. M.L. Mpampa, P.N. Azariadis, N.S. Sapidis, A new methodology for the development of sizing systems for the mass customization of garments. Int. J. Cloth. Sci. Technol. **22**(1), 49–68 (2010). https://doi.org/10.1108/09556221011008802
35. G. Cooklin, *Pattern Grading for Women's Clothes: The Technology of Sizing* (Blackwell Science, Oxford, 1999), pp.3–18
36. D. Gupta, B.R. Gangadhar, A statistical model for developing body size charts for garments. Int. J. Garments Sci. Technol. **16**(5), 458–469 (2004)
37. prEN 13402-3, Size designation of clothes—Part 3: measurements and intervals, Final Draft (2004)
38. X. Sibei, *Sizing System Created Using Size USA Data for Three Body Shapes* (North Carolina State University, North Carolina, 2013)
39. ASTM. ASTM international—about ASTM international. http://www.astm.org/ABOUT/overview.html

40. B. Hamad, S. Thomassey, P. Bruniaux, A method to reduce the number of sizes in apparel industry, in *IOP Conference Series: Materials Science and Engineering* (Institute of Physics Publishing), June 2020. https://doi.org/10.1088/1757-899X/827/1/012031
41. O. Ola-Afolayan, P.E. Zwane, A. Mastamet-Mason, Statistical modelling of key body dimensions in developing the size chart for the south African pear-shaped women. J. Cons. Sci. **49**, 2021
42. M.M. Majidi, M.B. Moghadam, G. Nourani, Pourtaheri, Development of sizing system for Iranian high school male students. World Appl. Sci. J. **13**(8), 1807–1813 (2011)
43. A. Destaw, Development of anthropometric size charts for Ethiopian Amhara region girls ages 5–10 years for better fitting garments. [Online]. Available: https://ssrn.com/abstract=3618893
44. S. Watkins, *Clothing. The Portable Environment*, 2nd ed (Iowa State University Press, Ames, Iowa, 1995)
45. C. Istook, S.J. Hwang, 3-D body scanning systems with applications to the apparel industry. J. Fash. Mark. Manag. **5**(2), 120–132 (2001)
46. L. McKinnon, C. Istook, Comparative analysis of the image twin system and the 3T6 body scanner. J. Text. Apparel, Technol. Manage. **1**(2), 1–7 (2001)

Chapter 4
Introduction to Pattern Making

Madeha Jabbar and Saeed Akhtar

Abstract Patterns serve as the blueprint for each and every article of clothing that people wear. Therefore, it is essential to have a fundamental understanding of the concepts that underlie pattern making. This chapter provides a comprehensive discussion of the process of pattern making. Guidance on how to construct essential blocks for basic garment components, how to choose the appropriate equipment and materials, and how to precisely measure one's body are included in this chapter. A brief history of the pattern making is also provided in the chapter to familiarize the readers with historical developments. A section has been dedicated to understanding of body measurements and sizing systems in apparel industry. The chapter is concluded with pattern grading and sizing including methods of grading and pattern adjustment for different body types. A careful reading of this chapter will help you acquiring the basic knowledge and skills required to translate ideas into wearable works.

Keywords Pattern making · Grading · Sizing system · Body measurements · Apparel

Introduction

The process of making patterns is an integral part of the garment manufacturing process. Patterns serve as the blueprint for each and every article of clothing that people wear. It is essential to have a fundamental understanding of the concepts that underlie pattern making, regardless of whether the objective is to create a fashion line or to personalize goods for one's own usage.

A comprehensive discussion of the process of pattern making has been presented in this chapter. Guidance on how to construct essential blocks for basic garment components, how to choose the appropriate equipment and materials, and how to precisely measure one's body are also included in it.

M. Jabbar (✉) · S. Akhtar
Department of Textile Engineering, School of Engineering and Technology, National Textile University, Faisalabad, Pakistan
e-mail: madehajabbar@ntu.edu.pk

M. Q. Khan et al. (eds.), *Garment Sizing and Pattern Making*, SDGs and Textiles,
https://doi.org/10.1007/978-981-97-7683-2_4

In order to create patterns, one must have a keen understanding of the dynamic relationship that exists between structure and design, shape and function, and technical precision. In order to ensure that clothing is not only aesthetically beautiful but also practical for the wearer, it is essential to investigate the concepts of proportion, balance, and fit [1]. A careful reading of this chapter will help you acquiring the basic knowledge and skills required to translate ideas into wearable works.

The creation of patterns is an essential component of the garment design discipline. The challenges associated with production and fit of apparel can be resolved by complicated and diverse process of pattern making. It is common practice to split pattern development into three distinct ways.

- Drafting.
- Flat pattern.
- Draping.

Drafting is a process that involves trial and error. Proportional coordinates are utilized in order to generate a basic pattern block that is straightforward. Creating a design that is formed in two dimensions is what is known as a flat pattern. There is also the possibility of modifying the basic pattern block in order to incorporate stylistic features. During the modeling or draping process, the fabric is draped directly onto the body or manikin. The application of this method is applicable to designs that are both straightforward and intricate. That being said, this is a difficult process that requires both knowledge and practice.

Depending on the size of the manufacturing, there are two different types of flat patterns. The first type of design is known as personalized design, and it involves adapting the pattern to the specific proportions and shape of an individual's body. The commercial design style is the second sort of design, and it is characterized by the use of traditional body forms and proportions. In order to generate efficient patterns for the garment that is supposed to be worn, it is possible to analyze the connection that exists between the domains of garment design, body measurements, and material. At this stage of the garment development process, the production of patterns is an extremely important step. When it comes to pattern creation, the two most crucial criteria are the dimensions of the user and the degree of ease allowance. When it comes to developing a basic pattern block for a static body, the process of basic pattern creation is driven by body measurements data. This is true regardless of whether the technique is carried out manually or through an automated CAD system. When it comes to the construction of patterns, both manual and through computer-aided design (CAD) require a wide variety of tools, including both hardware and software.

History of Apparel Pattern Making

Throughout the history of the garment manufacturing industry, the skill of pattern-making has always been absolutely essential to the success of the industry. A closer look at its historical growth shows information about the progression of technology, the effects it has had on society, and the ever-changing field of fashion.

Early Beginnings (Sixteenth–Eighteenth Centuries)

There are a few of notable publications around the sixteenth century in Spain, which can be considered as the earliest documented evidence of patternmaking. The first known instructional guides that were specifically devoted to the craft of pattern-making are Juan de Alcega's "Libro de Geometría, Práctica y Traca" (1589) [2] and La Rocha Burguen's "Geometrica y Traca" (1618) [3], as shown in Figs. 4.1 and 4.2 respectively. Both books were not only merely academic treatises, but they also acted as practical manuals for tailors, explaining in great detail the procedures that are involved in the process of developing patterns for clothing that are worn by people of different societies.

These early books were depended upon by tailors in order to transform their understanding of garment manufacturing into directions that could be followed.

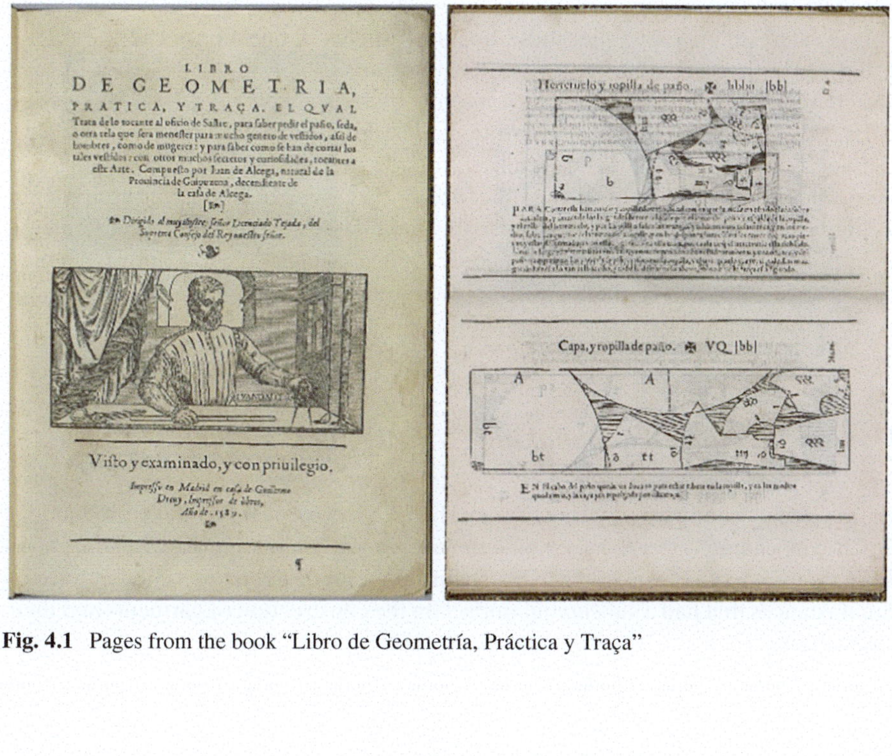

Fig. 4.1 Pages from the book "Libro de Geometría, Práctica y Traça"

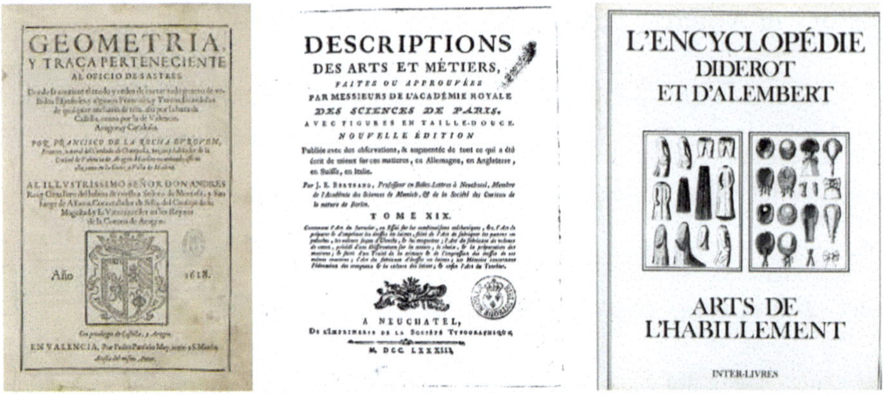
GEOMETRIA,
Y TRAÇA PERTENECIENTE
AL OFICIO DE SASTRES
Año 1618.
EN VALENCIA

DESCRIPTIONS
DES ARTS ET MÉTIERS,
FAITES OU APPROUVÉES
PAR MESSIEURS DE L'ACADÉMIE ROYALE
DES SCIENCES DE PARIS.
AVEC FIGURES EN TAILLE-DOUCE.
NOUVELLE ÉDITION
TOME XIX.
A NEUCHATEL,

L'ENCYCLOPÉDIE
DIDEROT
ET D'ALEMBERT
ARTS DE
L'HABILLEMENT
INTER-LIVRES

(a) (b) (c)

Fig. 4.2 **a** La Rocha Burguen's "Geometrica y Traca", **b** Garasault's "Descriptions des arts et mètiers", and **c** Diderot's "L'Encyclopédie Diderot et D'Alembert: arts de l'habillement"

Additionally, the books would provide thorough illustrations and step-by-step directions for drafting patterns for doublets, cloaks, hose, and other garments that were popular during that time period. These publications explained the geometric principles that underlie garment shapes. Consequently, this not only assured uniformity in the construction of garments, but it also made it easier for experienced tailors to pass on their skills to apprentices, so setting the groundwork for subsequent generations of patternmakers.

It is interesting to note that these Spanish articles in question were not unique instances. In the beginning of the eighteenth century, the Enlightenment in Europe was characterized by a resurgence of interest in the spread of practical knowledge and skills. Both Garasault's "Descriptions des arts et mètiers" [4] and Diderot's "L'Encyclopédie Diderot et D'Alembert: arts de l'habillement" [5] contributed to the further consolidation of the significance of patternmaking by offering in-depth sections that were specifically devoted to the trade. The concepts that were articulated in the previous Spanish works were not only replicated in these publications, but they also introduced new techniques and considerations, which reflected the changing world of fashion and garment manufacturing.

The Rise of Home Dressmaking (Nineteenth Century)

The nineteenth century was a time of great change in the world of clothing production, which was characterized by a considerable increase in the number of people who made their own dresses at home. This occurrence can be traced to a convergence of circumstances that had a substantial impact on the construction of garments and their accessibility.

The introduction of the sewing machine around the year 1825 was a significant step forward in the industry [6]. The garment assembly process was considerably expedited by this technological marvel, which resulted in the process becoming both more efficient and less labor-intensive. Individuals, particularly women, were given the ability to undertake sewing tasks at home as a result of this newly discovered simplicity of construction, which contributed to the development of a culture that values self-reliance and creative expression in terms of clothing choices.

The availability of flexible dress forms during this time period was a significant factor in the improvement of clothing wearability [7]. The more realistic draping and alterations that were possible with these forms, in contrast to their rigid predecessors, resulted in clothing that were better able to adhere to the specific body contours of each individual. This resulted in an overall improvement in the quality and comfort of clothing that was sewn at home, which further encouraged folks to engage in dressmaking initiatives themselves.

At long last, the widespread availability of low-cost paper throughout the nineteenth century made it possible to manufacture printed designs in large quantities without any difficulty [8]. The presence of this easily available resource eliminated the requirement for persons to create patterns from scratch, which resulted in a major reduction in the skill barrier involved with the manufacturing of garments. Another factor that contributed to the revolution in access to patterns was the establishment of commercial pattern firms. Not only did these businesses provide a wide variety of patterns, but they also pioneered the use of standardized size systems, thorough instructions, and even fashion magazines as promotional materials. By democratizing fashion and creating a thriving culture of home dressmaking, this all-encompassing strategy made it possible for individuals with varied degrees of sewing competence to participate in the fabrication of apparel.

Twentieth Century and Beyond, Innovation in Pattern Making

The field of patternmaking saw a period of exceptional advancement during the twentieth century. This progress was marked by the introduction of new techniques, expanding technologies, and an increasing emphasis on uniformity and efficiency. A three-pronged approach was used during this time period to revolutionize the process of garment creation. Each of these approaches made a substantial contribution to the development of the trade [9].

The use of standardized size standards emerged as a significant development in the early 20th century [10]. Before this, there was a lack of consistency in the sizing of garments among different manufacturers, which resulted in problems with the fit and discontent among customers. By using standardized size charts, individuals were able to purchase clothing with better confidence and reduce the need for extensive adjustments. This was made possible by the enhanced uniformity and predictability in garment fit that was caused by the deployment of these charts. This uniformity not only helped customers, but it also simplified the production procedures of clothing manufacturers, which led to increased productivity within the industry as a whole [11].

Another important development in the twentieth century was the introduction of new drafting processes, which allowed patternmakers to have a wider range of opportunities. In addition to more conventional techniques, such as sloper drafting, more contemporary approaches, such as flat pattern drafting [12] and draping on mannequins [13], were further utilized. In the book "Patternmaking History and Theory" by Jennifer Grayer Moore [14], the process of flat pattern drafting is described as the creation of two-dimensional representations of garment pieces. This method provides a greater degree of flexibility and control over the shaping process. The use of drapes on mannequins, on the other hand, makes it possible to manipulate fabric in three dimensions directly on a form. This allows for the visualization of the fit of the garment in real time and makes it easier to make alterations for a more customized fit. Patternmakers were able to manufacture clothes with high complexity and precision because of these innovative procedures, which allowed them to meet the ever-changing needs of the fashion industry.

Latter half of the twentieth century was the time when computer-aided design (CAD) software began to have a substantial and revolutionary impact. With the advent of computer-aided design (CAD), pattern production and manipulation underwent a revolutionary change, resulting in unparalleled levels of precision, efficiency, and flexibility. With the use of this technology, patternmakers are able to digitally draft designs, easily alter them, and generate automatic grading for a variety of sizes. The book written by Moore highlights the fact that computer-aided design (CAD) software not only simplifies the process of patternmaking but also makes it easier for organizations within the fashion industry to work together and share information, hence establishing a fashion ecosystem that is more effective and globally connected [15–17].

The twentieth century is a witness to the ongoing evolution of patternmaking, which has been going on for quite some time. This century witnessed innovations that not only improved garment fit and production efficiency, but also empowered patternmakers to push the boundaries of creativity and design. These advancements included standardized sizing methods, new drafting processes, and the transformative potential of computer-aided design (CAD).

Fundamentals of Pattern Making

The act of transferring three-dimensional structures onto a two-dimensional plane is the fundamental aspect of pattern making. This technique is referred to as flat patterning. Pattern makers are able to generate templates, known as "patterns," that serve as instructions for cutting and assembling fabric pieces. This is accomplished by breaking down particularly complicated shapes into more straightforward geometric forms. Having the ability to conceive in three dimensions is the most critical skill that a pattern creator may possess. Patternmakers, in contrast to traditional artists, who create exclusively in two-dimensional space, are required to visualize how flat pieces of fabric would drape and contour around the body. To accomplish this, a strong understanding of anatomy is required [18].

The notion of precision is one of the most important aspects of pattern design. When it comes to pattern making, exact measurements and calculations are required, in contrast to sketching or draping, which allow for a certain degree of interpretation. There is a difference between a garment that fits well and one that is badly built and that difference might be as small as a few millimeters.

Patternmakers utilize a wide range of equipment and methods in order to reach this level of precision in their work. There are a variety of tools that are utilized in the process of pattern design. These tools range from the simple tape measure to more specialized rulers and curves. Pattern designers, on the other hand, need to have a strong command of geometry, mathematics, and spatial thinking in order to be able to manipulate shapes and proportions with ease.

Understanding Body Measurements and Sizing

When it comes to successful pattern making, taking precise body measurements and having a solid understanding of garment sizing are two of the most important aspects. Both are extremely important in ensuring that a garment is well-fitting and flattering to the body shape of the person who is wearing it.

Body Measurements

Key measurements of body include essential measurements needed for drafting most basic garment patterns, and are shown in Fig. 4.3. They typically include:

- Hat size: Wrap a flexible measuring tape around the widest part of your head, keeping it slightly above your eyebrows. Ensure the tape is level and snug but not constricting.
- Neck circumference: Measure around the base of your neck, keeping the tape level and comfortable.
- Chest: Wrap the tape measure around the fullest part of your chest, keeping it level and parallel to the floor. Breathe normally while taking the measurement.
- Under chest: Measure around your torso directly under your bust, keeping the tape level and parallel to the floor.
- Waist: Measure around your natural waistline, where your body bends when leaning sideways. Keep the tape level and comfortable.
- Hips: Measure around the fullest part of your hips, keeping the tape level and parallel to the floor.

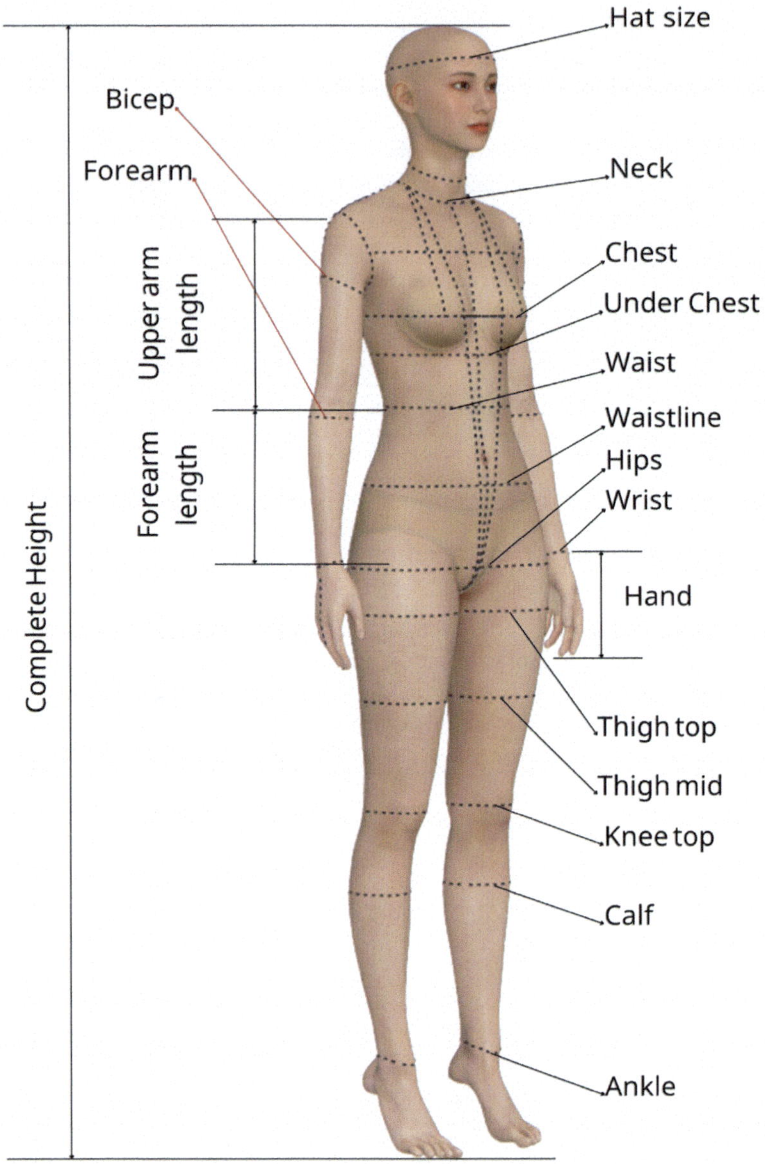

Fig. 4.3 Essential body measurements required for pattern making

- Bicep: Measure around the fullest part of your upper arm, keeping the tape level and parallel to the floor. Flex your arm slightly while taking the measurement.
- Forearm length: Measure from the elbow crease to the wrist bone, keeping your arm relaxed and slightly bent.
- Wrist circumference: Measure around your wrist bone, keeping the tape level and comfortable.
- Upper arm length: Measure from the shoulder point (where the arm meets the shoulder) to the elbow crease, keeping your arm relaxed and slightly bent.
- Thigh top: Measure around the fullest part of your thigh, keeping the tape level and parallel to the floor.
- Thigh mid: Measure around the middle of your thigh, halfway between the hip and knee, keeping the tape level and parallel to the floor.
- Knee top: Measure around the top of your kneecap, keeping the tape level and parallel to the floor.
- Calf: Measure around the fullest part of your calf, keeping the tape level and parallel to the floor.
- Ankle circumference: Measure around your anklebone, keeping the tape level and comfortable.
- Complete height: Stand straight against a wall, and measure from the top of your head to the floor.
- Hand length: Measure from the base of your palm to the tip of your middle finger, keeping your hand flat.
- Hand width: Measure across the widest part of your hand, excluding your thumb.

Taking Measurements

When it comes to patternmaking, taking precise measurements of the body is really necessary in order to get garments that fit well. In order to obtain accurate measurements, it is essential to maintain a comfortable and straight posture while wearing little clothing that does not affect the structure of your body. A flexible measuring tape needs to be utilized, and during the process, it should remain in a level and parallel position to the ground. It is important to maintain a relaxed posture and breathe normally while you are taking measurements. If you are measuring by yourself, it is recommended that you get assistance for places that are difficult to reach, such as the back and shoulders. Utilizing these procedures will result in the acquisition of trustworthy body measurements, which will serve as a firm foundation for the effective fabrication of patterns and garments.

Sizing Systems in Apparel

In the garment industry, standardized size methods are extremely important because they provide a common language for all parties involved, including consumers, retailers, and manufacturers. These systems classify persons into particular sizes based on a combination of body measures. The goal of these systems is to ensure that items from various brands and product lines fit in a manner that is consistent with one another. Nevertheless, establishing the complexities of standard size systems is a global difficulty due to the fact that there are inconsistencies and variations across a variety of countries and areas [11].

Common Sizing Systems

Standardized sizing systems including US, UK, and EU sizes are the most common types of sizing systems. Generally speaking, the USA and Canada are the countries that make use of the US size system. This method classifies persons according to their bust, waist, and hip measurements. Typically, the range of sizes that are designated is from 0 to 32, with even numbers indicating greater sizes. The system used to determine sizes in the UK ranges from six to eighteen, with higher numbers signifying larger sizes.

A method that is distinct from both the USA and the UK's systems is utilized by the European Union (EU) size system, which is utilized everywhere in Europe. The system employs a numbering system that ranges from 32 to 54, with higher numbers denoting larger sizes.

The petite and junior sizing options that are offered by certain companies are designed to accommodate persons who have shorter torsos or thinner bodies, respectively. In some cases, these groups may have their own size ranges that may not immediately correlate to the standard sizing specifications. In contrast to the systems used in the USA and the UK, the European Union system offers fewer alternatives for petite and junior sizes.

Even though there are national guidelines, individual brands frequently create their own size charts depending on the preferences of their target audience and how they like their products to fit. The result of this is that the sizing of different companies might be inconsistent with one another, even for items that are labeled with the same size.

Vanity Sizing

In the apparel industry, the practice of vanity sizing adds another layer of complexity to the process of determining the appropriate size and fitting. There is a possibility that brands will modify their size charts in order to make clothing appear smaller than their actual measurements. This could potentially mislead customers and lead to frustration about their expectations for an appropriate fit. The apparel industry is therefore subject to a skewed reality that is presented by vanity sizing. There is another term for vanity sizing, which is size inflation.

Vanity size can be motivated by a number of factors, including psychological attractiveness, pressure from competitors, and the utilization of marketing methods to increase sales. Manufacturers are under the impression that labeling goods with smaller sizes will result in an increase in sales since consumers frequently link smaller sizes with being leaner and more appealing. Brands may resort to vanity sizing in order to appear more appealing to customers who are looking for smaller sizes in a market that is extremely competitive. This may occur even if the garments that the company sells fit similarly to those that are sold by competitors that have larger size labels. As a marketing tactic, smaller size labels can be used to create the impression of exclusivity or attractiveness, which has the potential to influence the purchasing decisions of consumers [19].

Consumers are left with a poor body image and confusion as a result of vanity sizing, which in turn leads to a decrease in their faith in brands. Consumers who have difficulty finding garments that fit them well may experience confusion and irritation as a result of the inconsistency that exists between the size that is labeled and the actual measurements of the garment. In particular, persons who do not fit into the range of sizes that are considered to be ideal can be more likely to experience negative body image and dissatisfaction as a result of the focus placed on smaller sizes. Vanity size can cause consumers to face frequent fit concerns, which can lead to a loss of faith in businesses and a reluctance to purchase apparel from unfamiliar merchants or websites.

Additionally, it has an effect on the environment since oversized items that are labeled with smaller sizes might result in an increase in the number of returns and waste resources, which contributes to the environmental impact that the fashion industry also has. It may be possible to solve the growing problem of vanity sizing by promoting consumer awareness campaigns, putting an emphasis on body positivity, and implementing industry-wide standards.

Even if these standardized solutions are available, it is still difficult to find a solution that is truly universally applicable. There are a number of reasons that contribute to this difficulty. Even among people who fall into the same size group, there can be considerable variances in the proportions of their bodies (for example, having a long torso rather than short legs). When it comes to a particular size, this can result in problems with the fit for certain body shapes. While the methods used in the USA, the UK, and the European Union are the most prevalent, there are numerous more regional variants that further complicate the issue of size consistency across international markets.

Understanding Ease and Ease Allowance

Ease, which is often referred to as ease allowance, is a fundamental concept in patternmaking that plays an important part in determining how well a garment fits and how comfortable it is to wear. In this context, "sizing" refers to the intentional addition of additional space to a garment that exceeds the actual measures of the body. Because of this greater area, there is no restriction on movement. Being able to adjust to a range of body forms is essential. A garment that is easy to move and bend in allows for comfortable movement and bending without the garment feeling restricting or tight. Ease provides for some flexibility to accommodate these discrepancies and ensures a comfortable fit for a larger range of body types, which is beneficial for group of individuals who have diverse body shapes and proportions compared to one another. Different types of garments call for varying degrees of ease throughout the garment. As an illustration, a loose-fitting blouse will have greater ease than a garment that is more tailored. It is therefore beneficial when it comes to the draping and styling of the garment.

The term "ease allowance" can refer to two distinct categories: design ease and wearing ease. Creating gathers, pleats, or a relaxed silhouette are all examples of design ease, which is the ease that is introduced for specific stylistic goals. On the other hand, the term "earing ease" refers to the fundamental ease that is required for comfortable movement and the adaptation of many diverse body types.

In addition to the type of garment, the type of fabric, and personal tastes, there are a number of other elements that can influence ease allowance.

In general, dresses have a greater degree of ease than pants, and styles that are looser fitting tend to have a greater degree of ease than ones that are more fitted. Fabrics that are woven and have very little to no flexibility require more ease than knit fabrics that have elasticity that is inherent to the cloth. When it comes to clothing, some people want a more relaxed fit, while others favor a more snug fit. At the time of selecting a size or modifying the ease permitted in a pattern, this individual preference must to be taken into consideration.

Comfort or static ease, dynamic ease, and style ease are the three categories that can be used to categorize ease allowances since they are based on the different types of physiological activities. It is essential to have static or comfort ease when doing fundamental activities like breathing while sitting or standing. Extreme postures, such as bending body parts, require dynamic ease in order to be performed properly. Style ease is required in order to generate a styled garment with greater ease in accordance with the desired shape.

It is possible for the degree of ease to be either negative or positive, depending on the function of the clothing, the properties of the fabric, and the perception of comfort that the user has. When creating a garment, it is possible to have better control over the garment's attractive fit if you treat ease as closely as you do body dimensions [20].

The Right Ease Allowance

The majority of commercial sewing patterns include recommended ease allowances for a variety of garment kinds and sizes inside their instructions. Depending on the preferences of the individual and the shape of their body, adjustments might be required. It is also possible to estimate the appropriate ease allowance by using fit samples. Before cutting into your final fabric, it is possible to experiment with different ease allowances and ensure a proper fit by creating a muslin mockup, which is a prototype garment created from inexpensive fabric. Even while body dimensions are quite important, it is equally important to take into account the ease allowed when selecting a size. It is possible that if you choose a size only based on your dimensions without taking ease into consideration, you will end up with a clothing that is either too tight or too restricting.

Fit Evaluation

As a crucial step in the production process, a fit evaluation is something that every industry is required to perform. One can evaluate the level of comfort provided by a garment in two different ways: subjectively and objectively or both. One's subjective evaluation can be influenced by the garment's fit as well as its overall feel. For the objective evaluation, on the other hand, the testing of fabric qualities is carried out in accordance with the protocols that have been defined. When conducting a fit evaluation, it is important to take into account both the psychological and physical components of comfort and pleasure. An effective method of fit testing can be utilized to bring about a reduction in the proportion of garments that are rejected. Because of this, it can be advantageous for the industry. The garment has complicated qualities, including the fit of the garment. Other methods that can be utilized to ascertain it include ease, grain, balance, and line. A garment's fit is determined by a number of factors, including and not limited to proportions, size, material qualities, body measurements, wearer's posture, garment drape, and garment design.

When clothing does not fit properly, it can hinder the flow of blood through the cardiovascular system, which can lead to skin irritation, uncomfortable moisture, and heat conditions, and ultimately, the wearer will experience an inconvenience. The fit of performance clothes ought to be improved. It is not acceptable for it to lower the work performances. In the process of bending and stretching, provide a more tailored fit and improved comfort, while also being simple to put on and take off.

Clothing fit depends upon:

1. Body dimensions and patterns.
2. Techniques of pattern making.

3. Body shapes.
4. Type of garment.

Tools and Materials for Pattern Making

For the patternmaker to be able to work effectively, they need to have the appropriate tools and supplies. In order to effectively communicate in the workroom and to reduce the number of errors that occur as a result of misunderstandings, the patternmaker should be familiar with and comprehend the language. In this chapter, we will discuss the many tools, supplies, and phrases that are utilized in the industrial sector. When they arrive at the project site, the professional patternmaker is equipped with all of the necessary instruments for patternmaking. There are a variety of places where one can obtain tools, including art stores, department stores, yardage stores, and fashion supply businesses. Following list of essential tools is shown in Fig. 4.4.

1. **Straight pins and pin holder**: Fine and sharp pins for draping and pinning patterns. Pin holder keeps pins organized and accessible (cushion or magnetic wrist/table).
2. **Scissors**: Paper for precise cutting, fabric for heavy-duty cuts without snagging.
3. **Pencils and pens**: Mechanical pencil for consistent lines, colored pencils for marking changes, felt-tip pens for pattern information.
4. **Rulers**: Flex general for short distances, 36-inch for longer lengths, plastic for curves, tailor's square for multi-function, triangle for right angles.
5. **Curve rulers**: French curve for armholes/necklines, sleigh curve for various curves, and hip curve for hiplines/hems/lapels vary form curve for blending curves.
6. **Hanger hooks/ringers**: Organize and store multiple pattern pieces together.
7. **Push pins**: Temporary holding of pattern pieces during manipulation and transferring muslin patterns.
8. **Stapler and remover**: Prevents pattern shifting when stapling multiple paper layers, easy removal without damage.
9. **Magic mend scotch tape**: Mends minor tears/rips in patterns temporarily.
10. **Notcher**: Creates notches for seam allowances, centerlines, ease, and front/back identification.
11. **Tracing wheels**: Pointed for transferring shapes onto paper, blunted for use with carbon paper on muslin.
12. **Awl**: Creates small holes for marking dart ends, pockets, trims, and buttonholes.
13. **Metal weights**: Hold patterns flat and prevent shifting during tracing, marking, and cutting.
14. **Measuring tapes**: 60-inch for body measurements.
15. **Tailor's chalk** to create marks on seams and style lines.

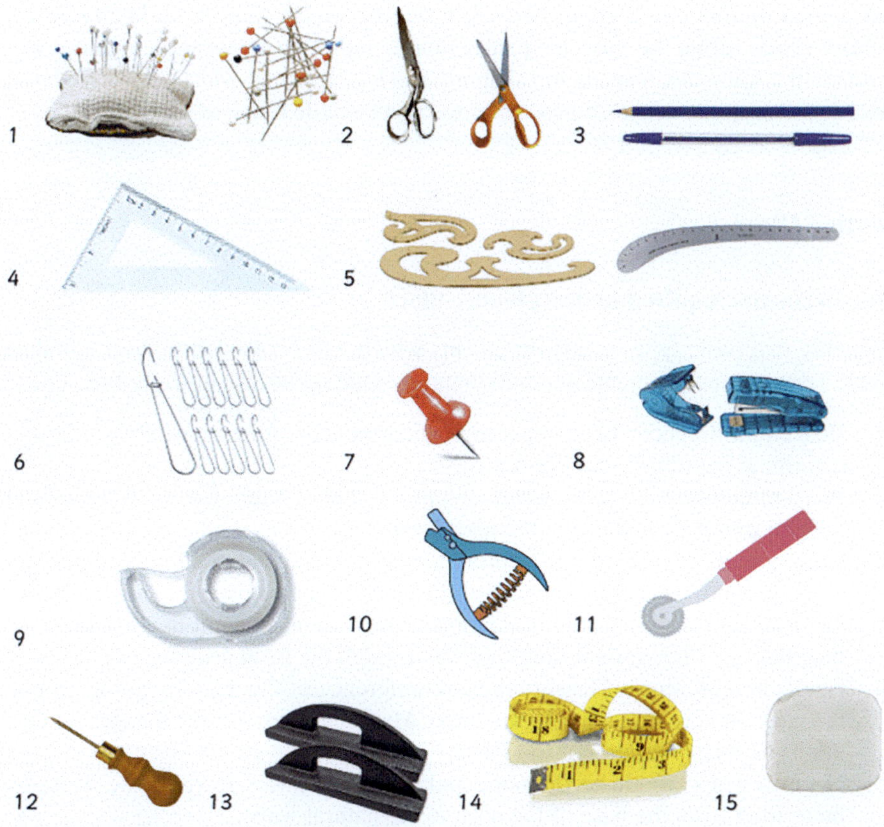

Fig. 4.4 Essential tools for pattern making

Drafting Basic Blocks

There are four basic blocks essential for learning the fundamentals of patternmaking:

1. Bodice block.
2. Skirt block.
3. Sleeve block.
4. Trouser block.

Bodice block forms the basis for tops, dresses, blouses, and jackets. It includes features like shoulders, armholes, bust darts, waist shaping, and a neckline. Skirt block serves as the foundation for various skirts, including straight skirts, flared skirts, and circular skirts. It establishes essential elements like waistline, hip circumference, and length. Understanding the skirt block enables modifications for different styles and volumes. Sleeve block shapes the sleeve component of garments. It incorporates elements like armhole opening, cap height, and ease for comfortable movement. We can create different sleeve styles like set-in sleeves, raglan sleeves, and bell sleeves. Trouser block forms the base for pants, shorts, and other legwear. It includes key features like waistline, inseam, hip shaping, and leg openings. Mastering these basic blocks and their variations empowers us to create a wide range of garment styles.

Bodice Block

Measurements required to draft bodice block

There are two basic types of measurements which will be needed to draft a bodice block: vertical measurements and horizontal measurements as shown in Fig. 4.5.

1. Bust circumference: Take a measurement around the fullest part of your bust to determine your bust circumference.
2. Waist circumference: To determine your waist circumference, measure around the part of your waist that is the narrowest.
3. Chest width: Keeping the measuring tape in a horizontal position with respect to the ground, measure the width of the chest by going over the bust and up to the armpits. From one side to the other, take a measurement that is exactly in the middle of the vertical lines that run through the front armpits.
4. Neck circumference: Measure the circumference of the neck by going around the neck and coming close to the torso. Maintain a grip on the measuring tape that is neither too tight nor too loose.
5. Nape to waist: Begin at the nape and work your way down to the waist by measuring from the base of the neck to the natural waist.

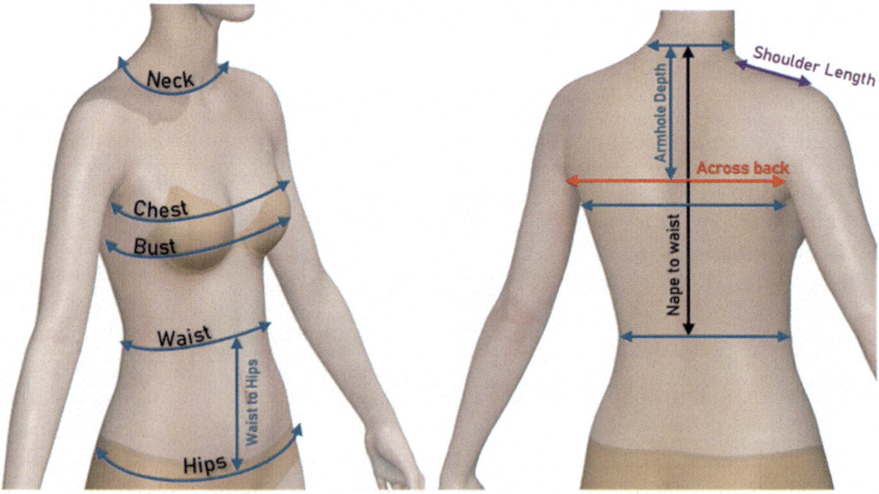

Fig. 4.5 Measurements required for bodice block

6. Shoulder length: Measure the length of your shoulders by starting at the base of your neck and going all the way down to the point where the shoulder joint moves when you lift your arm.
7. Back width: Back width is a horizontal measurement of your back that is started at the bottom of a standard armhole and extends all the way across to the other armhole. At a distance of approximately 5.6 inches (14.5 cm) below the nape of the neck, the width of the back is measured horizontally across the shoulder blades.
8. Armhole depth: To determine the armhole depth, measure from the base of the neck at the back vertically up to a guideline that is measured two inches or five centimeters below the armpit. If you need assistance determining this guideline, you can use some elastic.
9. Hip circumference (optional): Measure the circumference of your hips by going around the fullest part of your hips. This step is optional. When it comes to drafting the bodice, you do not require this measurement. It is necessary if you intend to develop the draft into a top draft (for more information on this process, see the final steps in this lesson).
10. Waist to hip: To determine the waist-to-hip ratio, take a vertical measurement from the narrowest part of the waist to the widest part of the hips (this measurement is optional).

Drafting Front Bodice

- To begin outlining the pattern, mark point A on the paper, preferably in close proximity to the top right angle. Then, the measurement from the nape to the waist should be taken from point A down the line, and point B should be marked.
- From point A, extend three-fifths of an inch (1.5 cm) upward and mark point C. The measurement of the neckline depth for the back bodice pattern is going to be this (A-C), which is going to be our measurement.
- From point A down, apply the armhole depth plus one-fifth of an inch (half a centimeter), and then mark point D.
- The vertical side of the pattern has been outlined up until this point in time. Moving forward, let's proceed with the horizontal lines.
- Starting from point D, square up and apply half of the bust measurement in addition to the sewing ease, which in this particular example would be twenty-five centimeters or two inches. Take note of point E.
- In order to mark point F, you must first square up from point B, and then apply half of the bust measurement in addition to the sewing ease. Again, starting from point C, square up, apply half of the bust measurement plus the sewing ease, and mark point G. This will complete the process.
- Create a line that is straight and connects the points G, E, and F.
- Identify the lines C-B as "Center Back" and the lines G-F as "Center Front" in the diagram as shown in Fig. 4.6.

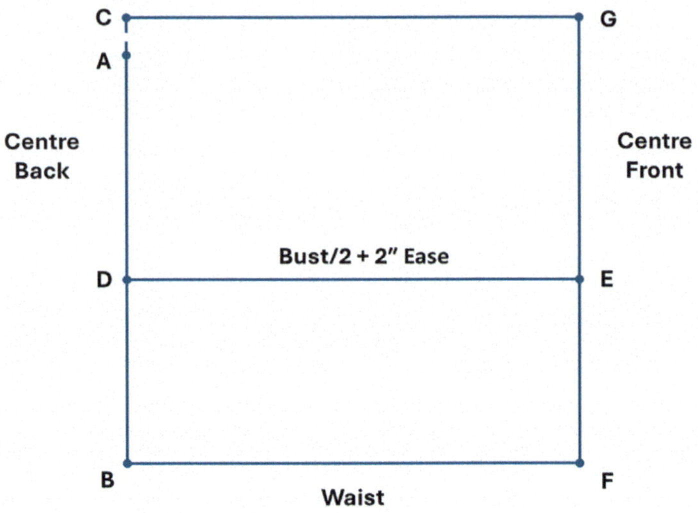

Fig. 4.6 Bodice block draft part-1

Drafting Back Bodice and Neckline

- Determine the armhole depth measurement, and then divide that number by 5. Deduct a quarter of an inch (half a centimeter) from this measurement.
- Make a mark at point H after you have measured down from point A and applied this number.
- Start at point H and draw a line through the area until you reach the line that connects F and G. To indicate the point of intersection, mark the H1 point.
- Take the measurement of the neck and divide it by five to determine the shape of the back neckline. Beginning at point C, use the 1/5 neck to measure and apply it in the direction of point G. Make a note of point I.
- Take the measurement of the neck and divide it by five to determine the shape of the front neckline. To this number, add up to 0.7 cm, or 1/4 inch. It is necessary to measure and apply this measurement beginning at point G and moving toward point C. Mark the point J.
- Take a measurement from point G and add one-fifth of the neck measurement to one-tenth of an inch (0.2 cm). Note the point K.
- Establish a connection between points A and I by means of a shallow curve that creates a 90-degree angle at the center back line. Through the utilization of the French curve, establish a smooth curve that connects points J and K, as shown in Fig. 4.7.

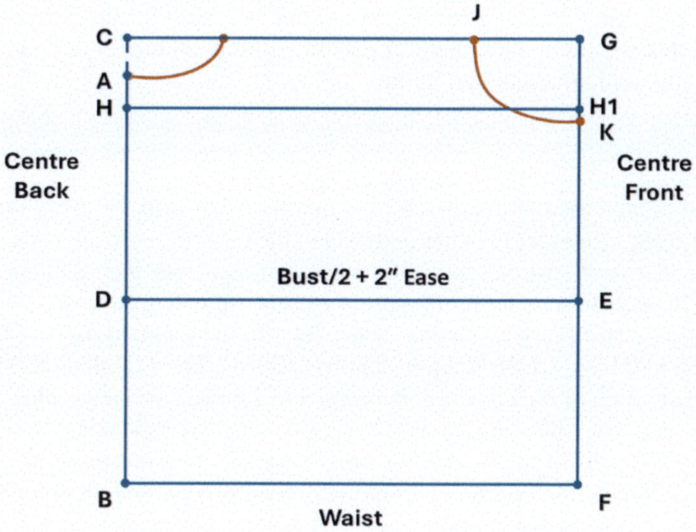

Fig. 4.7 Bodice block draft part-2

Drafting Back Shoulder

- Following that, you are going to require the measurement of the shoulder. The shoulder dart requires an additional 2/5 inch (1 cm) to be added to that measurement.
- Using the shoulder length plus two-fifths of an inch (one centimeter) measurement, start at point I. The beginning of the ruler should be placed at point I, and then the ruler should be moved upward until it reaches the shoulder length measurement, with the measure ending at the H-H1 line. Mark the L point. Establish a connection between point I and point L.
- The middle point between points I and L should be measured. Mark the M point.
- Take a measurement of 2 inches (5 cm) down from point M and 2/5 inches (1 cm) to the left. Identify this point as N. The point of the shoulder dart is denoted by this. 1/5 inch (0.5 cm) should be marked on each side of point M. To create the dart legs, draw a line from point N to each of these points along the length of the line.

Drafting Front Shoulder

- It is necessary for us to take two measurements in order to draft the front shoulder. First one is the length of the shoulder, and the width of the shoulder dart is the second factor.
- In order to accommodate the proportions of the body that are growing, the front dart expands with each increment of the bust, and the size of the dart is determined by the measurement of the bust.

- A bust measuring 34 inches or 88 cm is used as a base for the pattern-drafting method that we are utilizing here. A dart measuring 2½ inches or 7 cm in width is typically used on a standard figure.

The size of the dart then increases or decreases in proportion to the size of the bust in the following manner:

- The dart is increased or decreased by a quarter of an inch for every two inches after 34 inches. In order to better understand the formula, here are two examples.
- Assume that your bust measures 38 inches, which is four inches higher than 34 inches. In order to account for these four inches, we will need to add a quarter of an inch twice (for every two inches above 34). To the standard dart of 2½ inches, we add a half inch, which is equivalent to twice a quarter of an inch. This gives us the dart width of 3 inches, which is equal to 2 and a half inches plus one inch.
- Assume that your bust measures 32 inches, which is two inches less than 34 inches. The result of this is that we need to deduct one-fourth of an inch from these two inches. The width of the dart is determined by subtracting ¼ from 2½, which equals 2¼ inches as shown in Fig. 4.8.

- Using point E as a starting point, measure and mark half of the chest measurement in addition to one-half of the DART size, and then mark point O to complete the measurement. Find the point that is exactly in the middle of E and O, and mark it as P.
- Mark the point Q after you have squared down from point P toward the waistline. A dashed line should be used to connect P and Q.

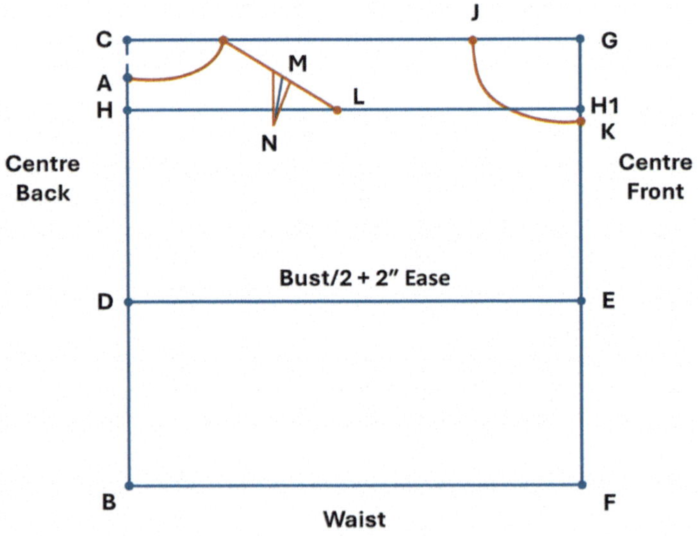

Fig. 4.8 Bodice block draft part-3

- Mark the bust point (BP) by measuring one inch (2.5 cm) down from point P. This will allow you to determine the bust point.
- The shoulder dart width should be measured and applied from point J toward point C, and then point R should be marked.
- Make a connection between each of the points J and R and the bust point BP in order to form the dart.
- With the point O as your starting point, draw a guideline that extends all the way to the H line.
- It is necessary for us to draft one more guideline in order to complete the shoulder. At a distance of three-fifths of an inch (1.5 cm) from the H-H line, draw a new horizontal line that runs across the block.
- Determine the shoulder length by beginning at point R and working your way outward. Place the beginning of the ruler at point R, and then move the ruler upward until you reach the measurement of the shoulder length. Finally, finish at the guideline that we have just drafted. Mark the S point. This will connect point R with point S.
- Beginning at point D, measure and apply a measurement of half across the back, in addition to a measurement of one-fifth (0.5 cm) on the bust line. Mark the T point.
- From point T, square up and draw a line until you reach the point where line H-H1 intersects. It is the point U. Locate the point that is exactly in the middle of D and T, and then mark it as point V. Mark the point W after you have squared down from the V toward the waistline. A dashed line should be used to connect V and W as shown in Fig. 4.9.

Drafting Armholes

Moving on to the armhole is the next step. From point O, measure up and apply one-third of the armhole depth measurement to mark point X. This should be done starting from point O.

- After locating the point that is exactly in the middle of U and T, label it as point Y.
- Find the point that is exactly in the middle of the T and O points, and then mark it as Z. In the direction of the waistline, descend in a square from Z, and mark the point Z1. A dashed line should be used to connect Z and Z1.
- Draw a diagonal line that is 45° inward and is 1 inch (2.5 cm) long, beginning at point T, and mark the point that corresponds to it as T1. When you begin at point O, draw a diagonal line that is 45° and is 4/5 inches (2 cm) long. Mark the point O1 on this line.
- Lines that are straight should be used to connect the points S, X, O1, Z, T1, and Y as shown in Fig. 4.10.

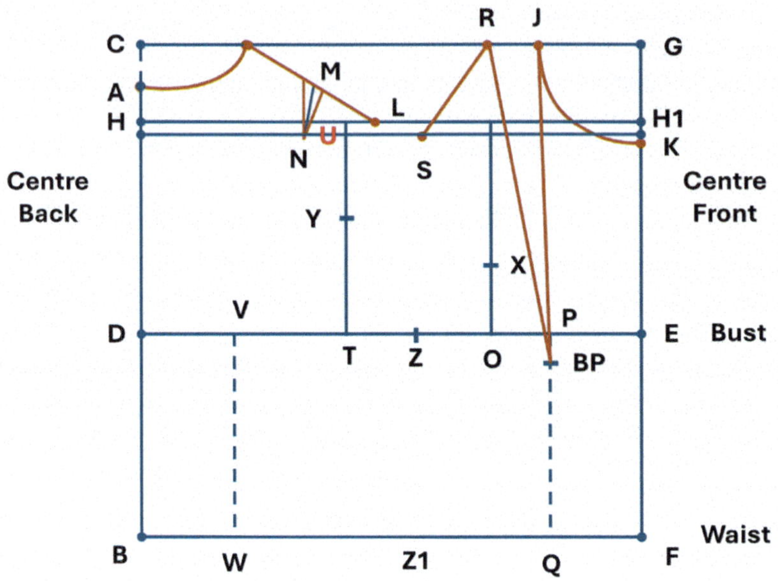

Fig. 4.9 Bodice block draft part-4

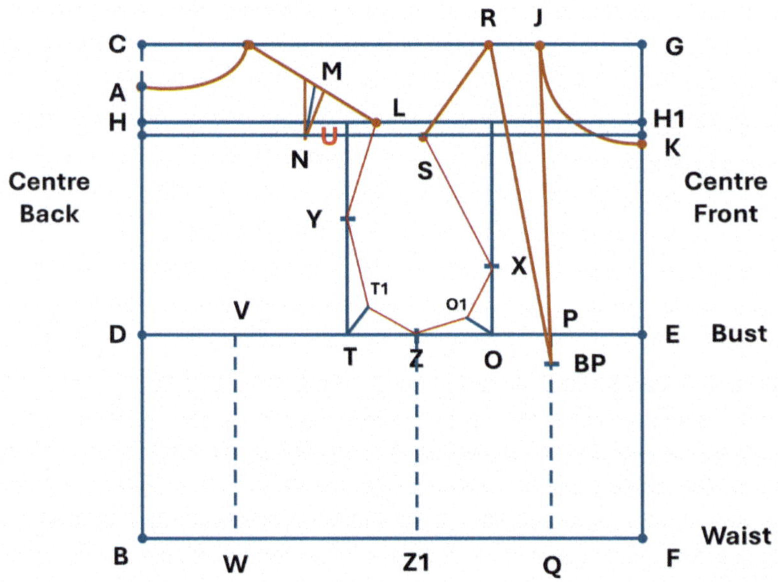

Fig. 4.10 Bodice block draft part-5

Drafting Side Seams and Waist Darts

- Extending two-fifths of an inch, or one centimeter, in a straight line from point F will ensure that the waistline will remain straight and will not ride up as a result of the bust. Make a connection between this extended point and place B.
- On average, the circumference of our bust is greater than the circumference of our waist. Addition of darts is required in order to form the waist. One dart will be placed at the front bodice, and the other dart will be placed at the back of the bodice. A dart-like drafting technique will also be used for the side seams.
- Through the course of the bodice pattern-drafting tutorial, a step-by-step demonstration of drawing precise pattern lines on paper is provided.
- First things first: Let's determine the size of each waist dart.
- Please take a measurement of your waist. In order to maintain uniformity, it is necessary to incorporate a two-and-a-half inch (or six centimeter) ease into the waist, just as we have done with the bust.
- In order to construct the pattern for half of the body, which is comprised of a quarter of the front and a quarter of the back, you must first determine your waist measurement and then divide it by two. Take this measurement and subtract it from the measurement of the half bust plus ease. When it comes to each of the three waist darts, divide the total number by three.
- Each third should be distributed in an even manner on both sides of the points W, Z1, and Q. In the event that you obtain a dart measuring 1 inch (2.5 cm), for instance, you should mark a half inch (1.25 cm) on each side of the points W, Z1, and Q. The dart legs should begin at these marks, which serve as starting points.
- For the purpose of forming the front dart, connect the dart leg points around point W with point V, the dart leg points around point Z1 with point Z (these are actually the side seams), and the dart leg points around point Q with the bust point BP.
- If there is not much of a difference between the measurement of the bust and the measurement of the waist, the darts may be smaller or may not be required, depending on the circumstances. In future, when you are fitting the muslin, you will be able to figure this out as shown in Fig. 4.11.

Finalizing the draft

- We are almost finished with the basic bodice draft, but there are some finishing touches that need to be made. A French curve should be used to smooth out the armhole lines after they have been crossed. The pattern should be outlined, with your marks, notches, and seam allowance being added as shown in Fig. 4.12.

Pattern Grading and Sizing

Grading is the process of creating different sizes of a garment while maintaining its proportions. It starts with a basic size, which could be any size used for the initial pattern. The aim is to ensure that when a garment is scaled up or down, its shape remains consistent.

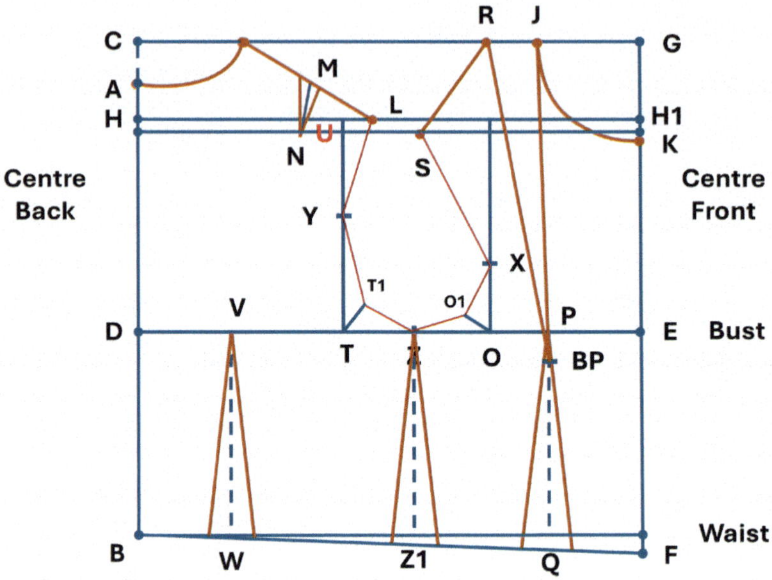

Fig. 4.11 Bodice block draft part-6

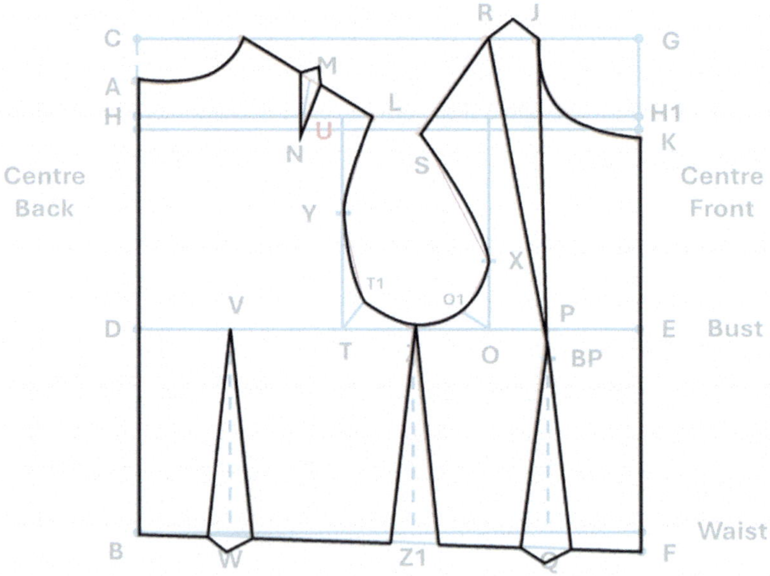

Fig. 4.12 Complete bodice block outlined

In the nineteenth century, the science of grading was created due to mass production and the introduction of commercial patterns. A pattern required grading to accommodate various body sizes and shapes. Currently, garment manufacturers and patternmakers utilize a medium-sized pattern (usually size S or M) as a foundation and adjust it for larger or smaller sizes through grading.

In grading, standard measurements are used, but it's important to note that not everyone's body shape fits these standards perfectly. Grading ensures that as sizes increase or decrease, the garment remains well-fitted and comfortable.

There are several methods for grading patterns, each with its own advantages and challenges. These methods include the slash and spread method, grading by shifting, grading by the point, and others. Each method requires precision and attention to detail to achieve accurate results.

When grading patterns, it's crucial to maintain the integrity of the garment's design. This means ensuring that features like darts and seams are properly adjusted to fit each size. Additionally, understanding the principles of horizontal and vertical grading is essential for achieving balanced proportions across all sizes (Fig. 4.13).

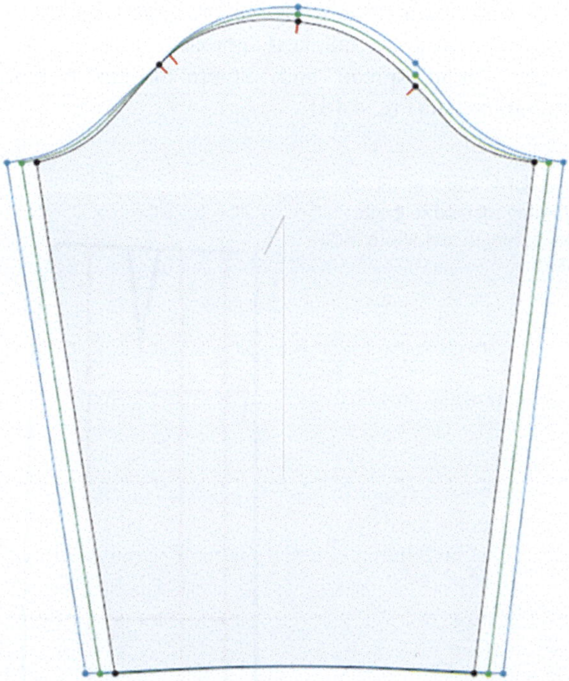

Fig. 4.13 Example of a graded pattern of a sleeve

Methods of Grading

There are two basic methods of pattern grading: manual and automatic. Manual grading involves cut and spread and pattern shifting method, while automatic grading is performed using CAD software.

Cut and Spread

This technique is regarded as the simplest. To grade up or down, this is accomplished by slicing the pattern and spreading the parts by a specified amount. Only a stylus, ruler, scissors, and measuring tape are required (Fig. 4.14).

Pattern Shifting

This technique involves shifting the pattern consistently at a fixed distance. The designer redraws the outline to achieve identical results to the previous cut and spread technique.

Computer Grading

Computer-aided grading has evolved alongside the advancement of new technology. It is regarded as the quickest approach. It essentially employs the same procedures as the "cut and spread" and "pattern shifting" techniques and converts them into a digital format (Fig. 4.15).

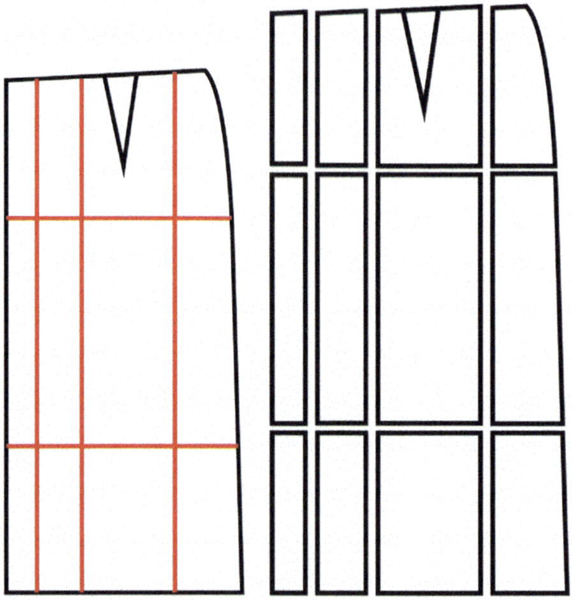

Fig. 4.14 Cut and spread grading method to grade along length and width for a skirt

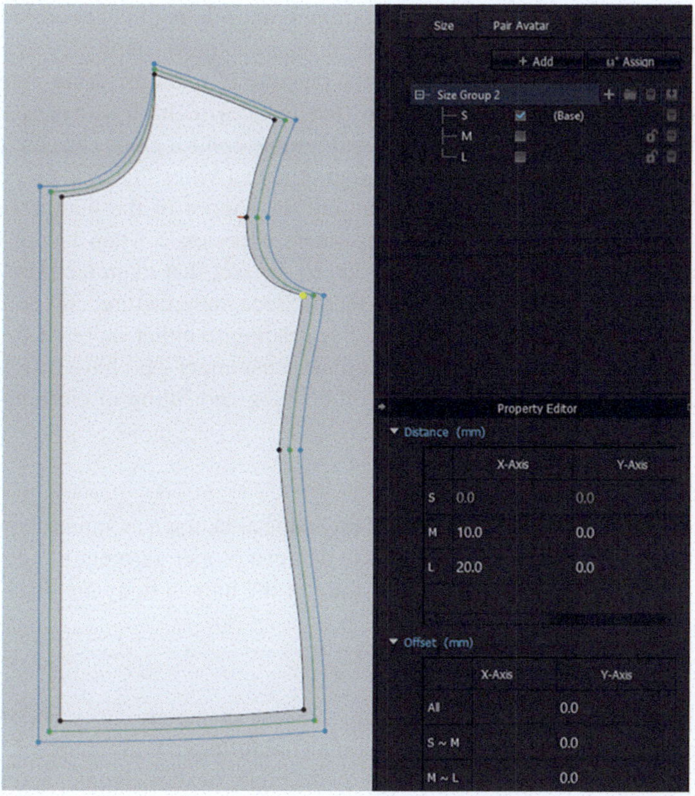

Fig. 4.15 Example of computer-aided grading process using CLO3D

Pattern Adjustments for Different Body Types

Human Body Shapes

According to the fashion industry, the fit of garments and articles of clothing is determined by the body type of the wearer. Body types can be classified into three distinct shapes, which include geometric shapes, shapes that are defined by fruit names, and other shapes that are referred to as "shaped things." Oval, square, triangle, rectangle, and circle shapes are those that are further subdivided into the category of geometric shapes. Apples and pears are the two types of shapes that are classified according to the names of the fruits. The following are some examples of other shapes that can be defined as shaped things: bean shapes, hourglass shapes, and heart shapes. These groups offer only a limited amount of information regarding the adaptation of these shapes to clothing. When it comes to the complaints that customers have about garments not fitting properly according to their body shapes, there are two components at play. Because clothing brands manufacture garments, the first one is for essential body shapes. This is because clothing brands do not manufacture garments for individual sizes. A significant number of manufacturers employ professional fit models who possess the ideal shape for the purpose of manufacturing garments for both males and females. Everyone is aware that the rates of obesity are increasing, that the consumption of fast food, computer, and remote control games has an impact on physical exercise such as yoga, jogging, and other activities that take place outside, and that these activities are rapidly changing the shapes of the human body. The second factor is the frustration that consumers experience when trying to find a garment that fits them properly. This is due to the fact that manufacturers produce garments in a variety of sizes by grading them. Since manufacturers do not have any information regarding body shapes, they use grading to either increase or decrease the size of the base pattern. On the other hand, customers who have different body shapes often express dissatisfaction with the sizing and fitting of clothing because the garment does not fit them properly.

Somatotyping

The shape of the human body and type of physique are known as somatotype. Nowadays, clothing industries take an interest to achieve proper garment fit according to the body types [21]. The shape and physique of the human body are classified into three categories shown in Fig. 4.16.

Women's Body Shapes

Women's body is in different sizes and shapes. Some shapes are curvy, and some have broader shoulders and tight hips. Every woman has different body shapes. Categories of women's body shapes include: hourglass, triangle, oval, rectangle, and diamond. The most common women body shapes are: hourglass, rectangle, inverted triangle/apple, and triangle/pear. These shapes have a little bit of difference. Within one type

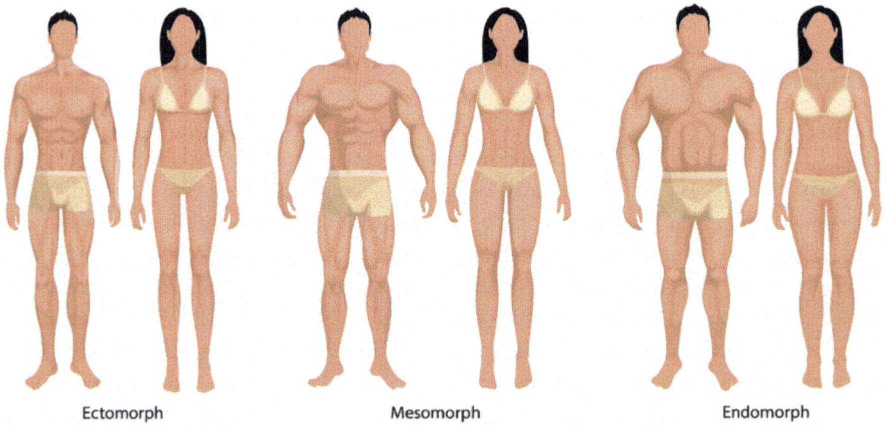

Fig. 4.16 Categories of somatotypes

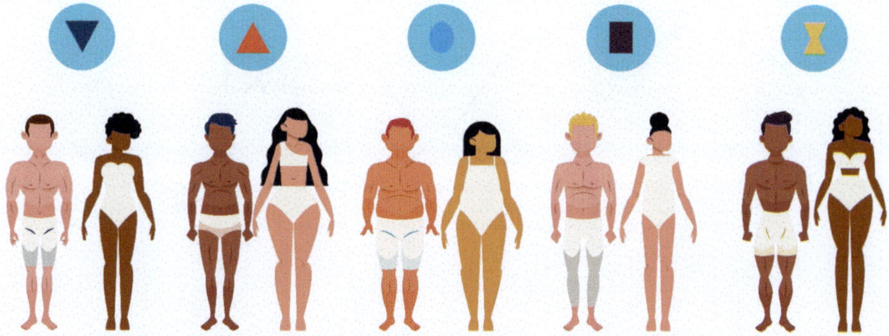

Fig. 4.17 Men's and women's body shapes

of body shape, there is a lot of variations. Different females' body shapes are shown in Fig. 4.17.

Men's Body Shapes

According to the age, shapes of men's bodies change. The human body is comprised fat, water, muscles, and bones. According to age change, people gain or lose weight, which can change body shape. As well as, consumers' need for clothing fit changes according to their body shape. Body shape also affects the fit preferences of consumers. Most frequently, the bodies of women are represented in media than men's bodies. Over the last ten years, sociologists and psychologists interested to study the satisfaction of men's body shapes. Most of the males like the muscular type of body.

By the use of somatotype of men's body shapes can be classified into geometric shapes which compare with vegetables, fruits, and other objects. Categories of men's body shapes could help the clothing industry to develop fitted garments according to the body shape and size of the target market.

Finalizing Patterns and Creating Technical Drawings

This process involves making adjustments to the initial pattern that was created based on the design that was selected and the measurements of the body. Generally speaking, it entails the following steps:

In order to evaluate the pattern's fit, comfort, and ease of movement, it is first tested on a body form or a prototype made of muslin. The evaluation is used to inform the adjustments that are made. Adjustments of a minor nature are made in order to smooth out lines and curves and to guarantee that the various pattern pieces are aligned correctly.

When stitching and finishing the pattern pieces, seam allowances are included around the edges of the pattern pieces to account for these processes. The construction

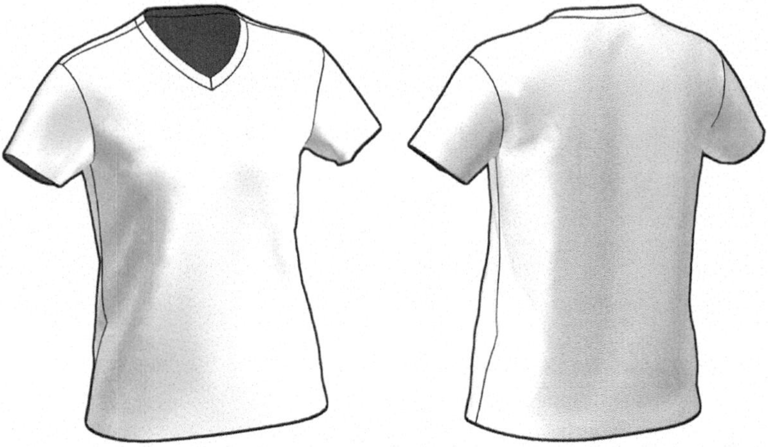

Fig. 4.18 Technical front and technical back of a shirt

process is facilitated by the addition of notches, which serve to facilitate the alignment of various pattern pieces. The use of darts helps to create shape and improves the fit of the garment. Every piece of the pattern is clearly labeled with its name, size, and grain line, which refers to the direction in which the fabric thread is running.

Following the completion of the final pattern, technical drawings are drafted in order to effectively convey the technical aspects of the design. These drawings act as a form of blueprint for the construction of garments and are essential for the processes involved in production. An example is shown in Fig. 4.18.

The front and back views of the garment are included in the flat sketches, along with annotations that indicate seam lines, darts, pockets, and other details concerning the garment. Included in this package is a list of materials, quantities, and construction details, along with detailed instructions. The term "bill of materials" (BOM) refers to the process of compiling a comprehensive inventory of all the materials that are necessary for the production process. This inventory includes the type of fabric, the quantity, and any additional notions such as zippers or buttons.

Summary

From its beginning as a method of draping fabric directly on the body, the art of making patterns for clothing has evolved into a sophisticated practice that is significant not only historically but also technically at this point. An understanding of this historical development, that ranges from hand-drafted patterns to computer-aided design, provides context for the approach that is currently being utilized. Having a solid understanding of body measurements, sizing systems, and ease allowances is essential when it comes to designing garments that fit properly. For the purpose of achieving the desired fit, it is essential to possess essential tools and skills such as the ability to draft the basic bodice block, utilize grading techniques for different sizes, and adjust patterns to accommodate a variety of the body types. After the pattern has been finalized, clear technical drawings that include specifications and flat sketches are created. These drawings ensure that accurate communication and efficient garment production are achieved. The chapter equipped readers with the knowledge necessary to appreciate the intricate steps that transform fashion sketches into garments that fit properly through pattern making.

References

1. M. Bicho, M. Pereira, N. Belino, R. Miguel, Fashion design and fashion technology courses: practical skills in patterns making, garment manufacturing techniques, technologies, and applications for fashion industry. Human Fact. Apparel Text. Eng. **32**, 73 (2022)
2. J. de Alcega, Libro de Geometria, Practica y Traça el Cual Trata de lo Tocante al Oficio del Sastre (Editorial Maxtor, 2009)
3. N. Waugh, The cut of women's clothes: 1600–1930 (Routledge, 2013)
4. M. Daumas, R. Tresse, La" Description des Arts et Métiers" de l'Académie des Sciences et le sort de ses planches gravées en taille douce. Revue d'histoire des Sciences et de Leurs Appl. **7**(2), 163–171 (1954)
5. D. Diderot, J.L. d Alembert, J. Proust, L'encyclopédie Diderot et d'Alembert: arts de l'habillement (1989)
6. J. Parton, History of the Sewing-machine. Howe Machine Company, No. 38, N. Charles St., 1870.
7. J.L.C. Forbes, Development and evaluation of individualized instructional media for draping on the dress form (1978)
8. K. Siemensmeyer, B. Siegel, S. Ervine, J. Bullock, Solutions for digital textile printing, in *NIP & Digital Fabrication Conference*, vol. 15 (Society of Imaging Science and Technology, 1999), pp. 280–283
9. T.J. Kang, S. Min Kim, Development of three-dimensional apparel CAD system: part 1: flat garment pattern drafting system. Int. J. Cloth. Sci. Technol. **12**(1), 26–38 (2000)
10. S. Xia, C. Istook, A method to create body sizing systems. Cloth. Text. Res. J. **35**(4), 235–248 (2017)
11. S. Gill, A review of research and innovation in garment sizing, prototyping and fitting. Text. Prog. **47**(1), 1–85 (2015)
12. E.I. Ugwu, M.N. Ezeaku, B.I. Attah, U.M. Emeghebo, E.C. Eze, Application of computer aided design (CAD) and flat techniques in teaching pattern drafting by clothing lecturers in universities in South East, Nigeria. Int. J. Home Econ. Hosp. Allied Res. **2**(1), 29–43 (2023)
13. H.T.M. Nguyen, V.T. Ho, T.T. Hoang, Creating forms for women's clothing by draping techniques. VNUHCM J. Sci. Technol. Dev. **18**(2), 25–36 (2015)
14. J. Grayer Moore, *Patternmaking History and Theory* (2019)
15. Y. Cho, T. Komatsu, S. Inui, M. Takatera, Y. Shimizu, H. Park, Individual pattern making using computerized draping method for clothing. Text. Res. J. **76**(8), 646–654 (2006)
16. J.J. Fang, Y. Ding, Expert-based customized pattern-making automation: Part I. Basic patterns. Int. J. Cloth. Sci. Technol. **20**(1), 26–40 (2008)
17. K. Liu et al., 3D interactive garment pattern-making technology. Comput. Aided Des. **104**, 113–124 (2018)
18. C. Patel, Z. Liao, G. Pons-Moll, Tailornet: predicting clothing in 3d as a function of human pose, shape and garment style, in *Proceedings of the IEEE/CVF Conference on Computer Vision and Pattern Recognition* (2020), pp. 7365–7375

19. S. Ketron, K. Naletelich, Redefining vanity sizing: when bigger may be better. J. Fash. Mark. Manage: Int. J. **21**(4), 438–452 (2017)
20. Y. Sakata, K. Kim, M. Takatera, Suitable ranges of ease allowance for appearance of women's shirts of fit and straight styles. Int. J. Cloth. Sci. Technol. **33**(3), 448–464 (2021)
21. C.-Y. Chiu, R. Ciems, M. Thelwell, A. Bullas, S. Choppin, Estimating somatotype from a single-camera 3D body scanning system. Eur. J. Sport Sci. **22**(8), 1204–1210 (2022)

Chapter 5
Body Measurement Methods for Pattern Making

Zheng Gangqiang, Rida Alvi, Zafar Javed, and Gohar Ali Hayat

Abstract This article discusses which essential measurements of the body are required for garment pattern design: how to get such measurements, where to collect measurements and body information, as well as how to use them for pattern design and grading. Also, outline how scanner measurements will be utilized to inform pattern construction. Currently, pattern drafting is a manual process that can be done in a variety of ways. Pattern drawing methods have been evaluated to determine ease of use, comprehend principles utilized in the procedure, and even illustrate how to better recognize the particular size, form, and dimensions employed in the process. However, studies have been undertaken to determine how scanner technology significantly impacts modern pattern-making techniques. While the present research emphasizes whole-body drafting, the similarities between the approaches indicate that the findings will be applicable to pattern drafting strategies utilized in various apparel types. Assuming that all body scanner measurements are valid for garment product creation is incorrect. Body scanning is a huge advancement in capturing human measurements, which are crucial for product development. This research not only assesses the adequacy of scan measures for pattern production, but also suggests additional measurements to better match the wearer's size, shape, and proportions.

Keywords Body measurement · Measurement tools · 3D scanners · Fashion

---

Z. Gangqiang · G. A. Hayat (✉)
Department of Industrial Design, School of Art and Design, Wuhan University of Technology, Wuhan, China
e-mail: gohar.ali@iub.edu.pk

R. Alvi · Z. Javed
School of Art and Design, National Textile University Faisalabad, Faisalabad, Pakistan

M. Q. Khan et al. (eds.), *Garment Sizing and Pattern Making*, SDGs and Textiles,
https://doi.org/10.1007/978-981-97-7683-2_5

Introduction

The foundation of scales for measurement and guidelines has frequently developed; thus, there have now established methods for analyzing measurement variability and asserting measurement precision with a particular degree of accuracy. The measurement has developed into a systematic science with a strong theoretical and practical base [1].

To take body measurements, the individual who is being measured should be dressed only in underwear so that the measuring tool can lie flat against the body. The measures ought to avoid being taken in front of a reflecting surface because the client is likely to modify their stance, resulting in erroneous measurements.

Most studies have found that utilizing a certain pattern drafting process for different body shapes or sizes produces varying results. This emphasizes the need for differently shaped clothes to attain a good fit, as well as the importance of adopting a measurement system and pattern construction processes that better reflect individual or demographic diversity. Developing ways for gathering measures to inform pattern drafting would provide a strong basis and allow us to better understand the connections across one's body and a particular pattern. This link was entrenched in early tailoring texts, even though they had restricted measuring tools, but it is rarely clearly discussed in modern pattern-creating approaches, which have much-improved measurement technologies [2]. Body scanning enables us to collect data about the body that has previously been difficult to obtain. It allows us to investigate shapes, acquire new measurements, and define dimensions that were previously impossible with the present technology. Using 3D body scanning allows us to challenge established methodologies while also proposing new ways of designing patterns that are better informed by the body. This study will concentrate on manual body measurement and 3D scanner improvement [3]. Traditionally, garment patterns are created by manually measuring and applying proportions to produce a block [4].

Introducing more adequate measurement support pattern drafting approaches based on body scanning would greatly help garment design academics and practitioners. It allows them to better understand the body-pattern links. The time and cost of the several fits required to perfect a pattern for a garment could be minimized or even eliminated by implementing more accurate pattern-making methods, such as 3D body scanning technology and increased analysis [3].

Body Measurement and Its Anatomy

Tools Required for Getting Measurements

Length ruler or wall graph, curvature ruler (a tape designed with a hook and eye for marking the contour of the waist and weighing back and front length), conventional measure tape, small tailor's rectangular shape, carton strip, a pencil, remover, and measuring sheet (approx. 50 cm in length).

Additional Data for the Measurement Chart

Aside from the body measurement, additional suggestions on shape measurements, including an oval back and a tilting shoulder, are appropriate. It is encouraged that particular supplementary measures can be computed and compared to the measured ones to acquire data on what the customer wants in figure type (proportion variations) for pattern design. If variations arise, inspect the measures to eliminate measuring inaccuracies [5].

How Do You Take Your Body Measurements?

However, if you are purchasing current body data to create design patterns, you are probably going to be suited on suitable models; thus, you must be able to take measurements manually. First, ensure you're equipped with the appropriate tools.

- Use an adaptable tape to measure (longer is preferred for taller or larger sizes).
- Mark the natural waistline with a flexible measuring tape that includes hooks, eyes, snaps, string, or ribbon.
- Record and document measurements on a computerized or printed sheet.

You must additionally make sure that the individual you're assessing is wearing properly tailored undergarments, such as a lightweight, non-compressive set of tights and a light, non-compressive top shirt. If you're getting measurements for an individual outfit for a certain event (e.g., a formal costume, wedding clothes, formal attire, etc.), make sure that your model wears the same shoes and lingerie as the item of clothing, as these might impact the fit as mentioned in Fig. 5.1.

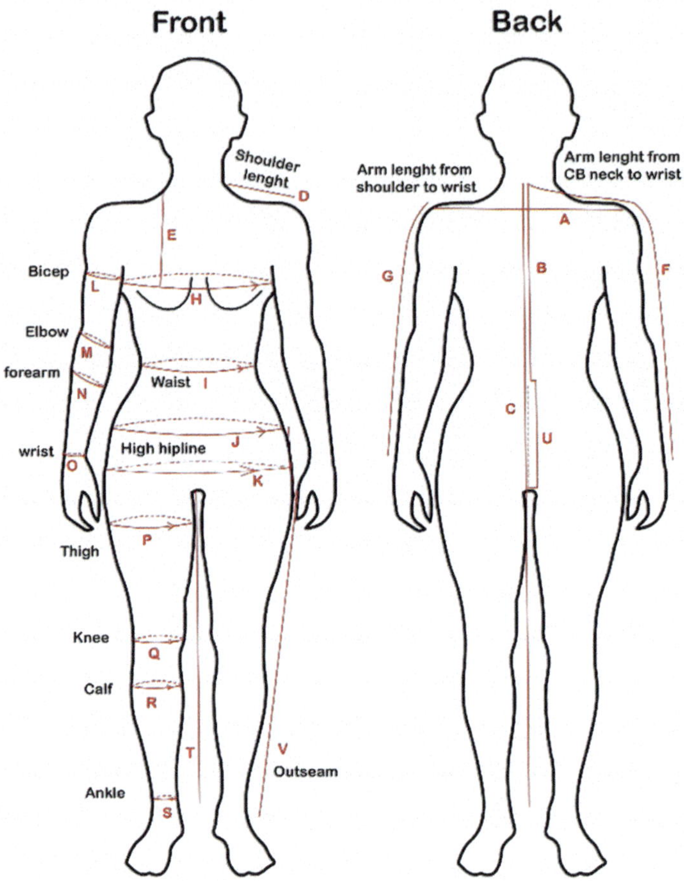

Fig. 5.1 Front and back measurements' points on body

Why Take Multiple Body Measurements for Different Dresses?

Design the prerequisites:

Inappropriate body measurements can hurt dress style. Even a slight short or long discrepancy might spoil the designs, making them appear sloppy and hazy. As a result, while creating clothing for a collection, provide enough space to change the fit.

Running Apparel:

The incorrect measurements might have an impact on the garment's quality and style. The garment creation (digital) technique, a long dress created for multiple occasions, is hampered by incorrect body measurements. Even with fitting clothing, incorrect dimensions can disrupt the design and appearance [6].

How Do You Take Significant Body Measures?

Full bust/chest—Measured from the broadest point of the bust/chest (often toward the top) around the entire body, parallel to the floor.

The full hip measurement is measured at the hip's fullest position (often at the bottom 1/3 of the buttocks), parallel to the ground.

Across the shoulder is the distance between the shoulder sockets. Sometimes elevating and then lowering the figure's arms will allow you to feel the socket joint.

HPS to apex is the distance between the HPS and the apex point.

CB—the vertical distance down the spine from the C7 vertebrae to the natural waist is known as the neck-to-waist measurement.

CF—the neck to waistline—the height between the base of the neck and the natural waist in the exact center front of the body.

Waist-to-hip length is measured down on the right side of the body, from the center of the waist to the actual hip position.

High thigh is the circumference of the human torso measured across the hip bones.

Waist circumference is determined around your waistline. It may be useful to have the client lean to the side to highlight their natural waist as shown in Fig. 5.2.

Total height—measure the second evaluate or thread around the wearer's actual waist. Extend from the CF waist position across the legs to the CB waist level.

Vertical trunk dimension spans from HPS, across the leading edge of the body, across the crotch, and up toward the back of the body to HPS as shown in Fig. 5.3.

Thigh: measure around the buttock's widest area. Approximately two or three inches under the hip area.

Inseam—the distance between the crotch and the floor on the inside of the leg as shown in Fig. 5.4.

What Exactly Would Be required for the Initial Drafting?

To begin with, flat pattern advancement requires a few items, including

> Include the measurement information for the size you would like, an outline or image of the outfit to be structured, and an illustration for common pattern angles and ratios.

Pattern-making procedures have advanced since the 1980s, with attempts to automate block sketching [7]. If you're designing your pattern on paper, you will require:

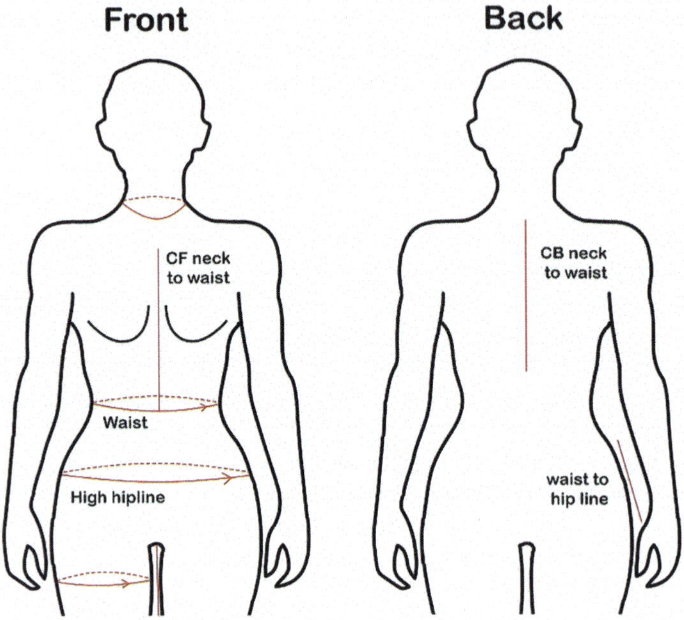

Fig. 5.2 Hipline, waist, and CB neck-to-waist measurements

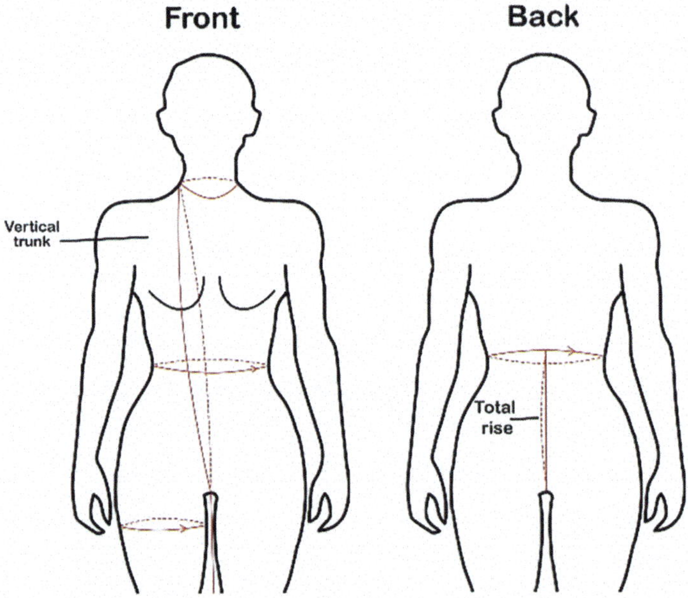

Fig. 5.3 Front and back vertical trunk and total rise measurement area

Fig. 5.4 Thigh and inseam measurements points

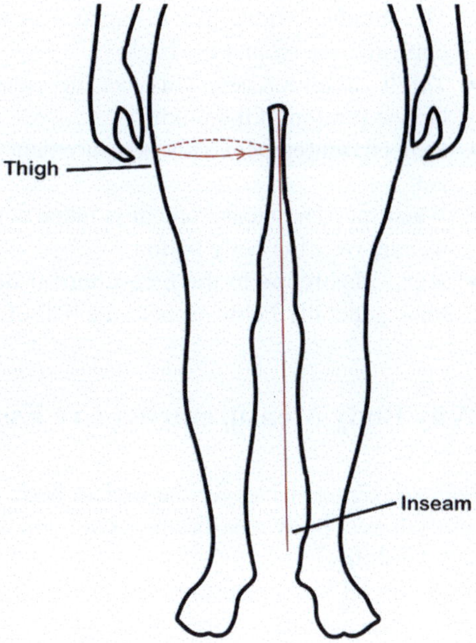

Pattern paper products, French curves, a flat ruler, and pencil.

Body Measurement and Pattern-Making Tools

Drafting can be done using a computer-aided design (CAD) application. We utilized Gerber Accumark, which you can see in multiple of the photographs, but alternative popular applications comprise Optitex and Lectra. Considering the reason why you're creating a pattern, and how you're doing it for oneself, a little enterprise, or a large organization, you'll have a clear idea of what your sense of style appears like. This may be an illustration of an actual garment, the illustration of the costume itself, a simplified sketch, or a tech pack comprising technical drawings.

Standard body measures vary slightly by country or location, but a few standard measurements involve:

- The height is the length from the lowest point to the top of the head when standing.
- Bust: This refers to a measurement determined around the widest part of the chest area, usually over the nipples for women and at the center of the bust for men.
- The Inseam: The measurement from the hip to the bottom of the heel bone while standing barefoot.
- The Arm length: The distance between the shoulder and the wrist bone.
- Neck measurement: This is the measurement that is acquired around the lowest point of the neck, commonly nearest to the collarbone.

- The shoulder width: The distance between the end of one shoulder bone and the opposite one around the back.
- Thigh circumference: This measurement is determined by measuring along the largest portion of the thigh.
- Calf circumference: This measurement is taken around the biggest area of the calf.
- Waistline: The measurement is taken across the narrowest area of the abdomen, usually over the belly button.
- Hips: This refers to the measurement determined around the widest part of the hips, generally below or near the ball of the hip bone.

Why Body Measurements Are Essential

Every designer strives for the perfect body measurement to complement the collection's theme. Body measurements are important for the dress's fit, style, appearance, and durability.

A correct body measurement aids in the maintenance of fabric quality through more effective fabric use. A single incorrect measurement disrupts the entire collection set, eventually affecting the fashion brand's value.

What Measurements Are Most Important for Pattern Making?

Our fundamental measurements vary based on the sort of garment you wish to construct. Let's start with some common acronyms. Although designers must define precise restrictions for the fabric quality, targeted age range for the collection, season, etc., here are a few guidelines that one should consider when practicing body measurement.

Bust—wrap the tape around the broadest part of your bust, but not too tightly or loosely. It will allow you to achieve the correct measurement. Cross shoulder—when obtaining an outfit stitched that will be worn on the upper body, one must have their cross-shoulder measurement taken. This is where the distance between one shoulder and the other is recorded. Armhole—measure from the outside edge of one shoulder to the armpits. Arm length—to measure the arm length, place a measuring tape from the armpit to the wrist. Wrist—measuring the circumference of the wrist's widest point would provide accurate measurements.

Waist—the portion directly above the belly button, with the smallest circumference.

Hip—to get the exact hip dimensions, measure the broadest area of the lower hip.

To measure the thighs, wrap the measuring tape across the largest portion of the thighs to make sure that the clothes fits properly. Calf—the measurement of the

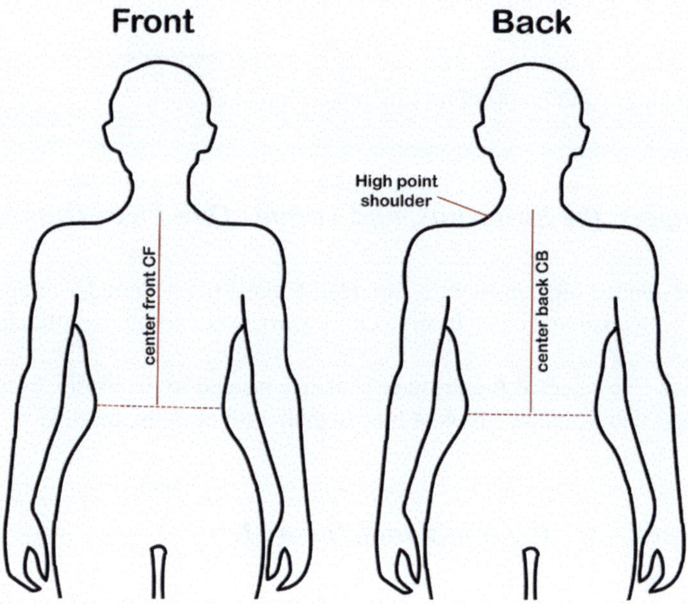

Fig. 5.5 CF, CB, and high point shoulder measurement points

calves is significant when making trousers or any other bottom wear. To get the right fit, measure the circumference around the calves.

Length—to determine the overall length of a dress, the model must stand straight. By then, the measuring tape should be at the top of the shoulder, descending to the floor or the desired length.

CB—center back: The center back of a garment or individual which runs along the vertebral column.

CF—center front: The area of a garment or body that extends from the flat at the middle of the neck down.

High point shoulder (HPS) is the most elevated point of the shoulder on the opposite side of the neck as shown in Fig. 5.5.

Blouses and Suits

- Measurements: full bust, full hips, across shoulders, highest point shoulder to apex, center back to neck to waist, center front neck to waistline, and waist-to-hip length. Dress dimensions include:
- Center back neck to bottom and center front neck to bottom.

Bottoms

Whole rise, thigh, and inseam Full hip, higher hip, and waist.

The Rompers, the Swimsuits, and Various One-Piece Outfits

Full bust, extended hips, crosswise shoulder, highest point shoulder to apex, center back neck to waistline, center front neck to waist, waist-to-hip length, and vertical column.

Finally, we've selected the primary statistics needed to determine every item in the category; therefore, let's look at how to gather all of these metrics.

What Exactly Are the Additional Torso Metrics?

Additional measurements of the body are still important in creating patterns, and further essential when evaluating your design afterward, thus getting comfy with them. They, like the key measurements, vary depending on the type of clothes for which you want to construct. The logic underlying the above is that if you're creating pants, you don't necessarily need to take a bust measurement, but you must still create a bra.

The Tops and Outfits

Measure your under bust, waistline, neck in the center, throughout the back and front, shoulder measurement, length of the arm from center back the CB neck to wrist, arm length from shoulder point to the wrist, the bicep, elbow, and forearm measurement.

Swimwear blouses

Front bust dimensions: apex to apex, cup width, cup depth, and apex to under bust.

Bottoms

Measure knee circumference, calf circumference, ankle circumference, and inseam/ out seam.

Additional Body Measurements: A Guide

The circumference of the vertebral column directly below the bust fold is referred to as the underbust.

Apex to the apex is determined from one top point to another with a straight tape to measure the surface of the floor as shown in Fig. 5.6.

The front bust is determined across the most natural side of the body at the top to the reverse side of the body. Top to underbust: determined at the highest point to the bust curve (Fig. 5.7).

The cup width is determined by measuring along the top of the cup from outside to inside the breast. The cup width is determined by measuring from the highest point of the bra to the chest fold, along the top edge. The neck basis is the diameter around the neck, determined along its bottom part as shown in Fig. 5.8.

Elbow length from shoulder point to wrist is measured from the shoulder socket to the outer line of the arm, next to the point of the wrist. The arms should be bent at the elbows.

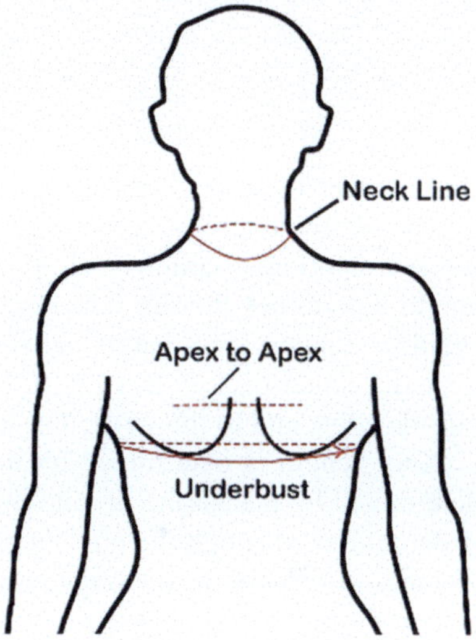

Fig. 5.6 Apex-to-apex and neckline measurements

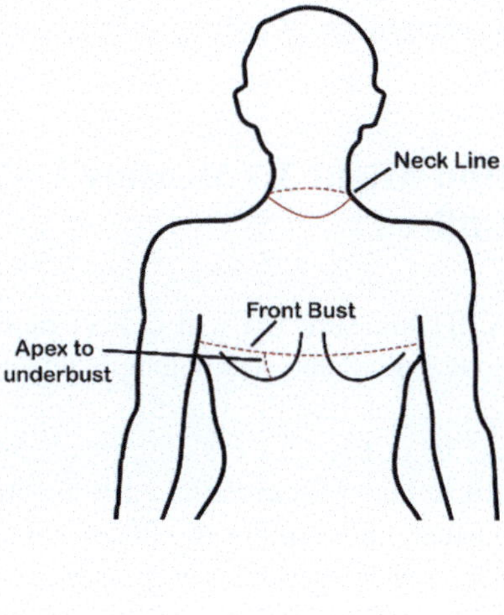

Fig. 5.7 Front bust and apex-to-underbust measurements' points

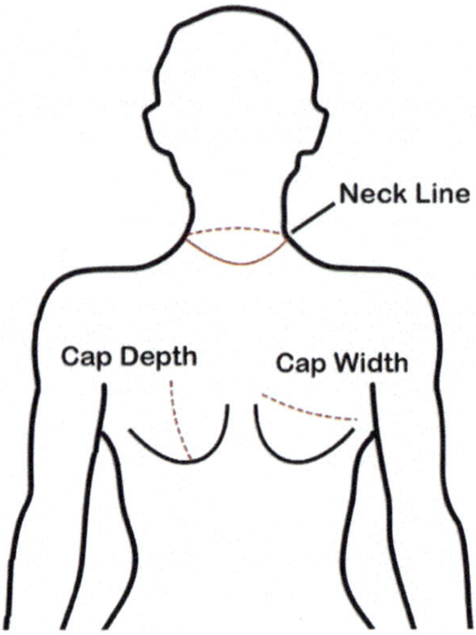

Fig. 5.8 Neckline, cup depth, and cup width measurements

The bicep diameter is determined by measuring at the widest part of the biceps.

The elbow length is determined at the elbow point. The circumference of the forearm at its longest point is referred to as the forearm circumference. The shoulders' length determines the distance between the highest point shoulder (HPS) and the shoulder column. Arm length between center back (CB) neck and cuff is calculated from the center back (CB) neck (C7 vertebrae) across the shoulder, along the arm's outside edge, and to the wrist. The arms should be bent little as shown in Fig. 5.9.

Knee length is determined by measuring within the kneecap. Calf diameter is determined by measuring around the longest part of the calf. Ankle circumference is determined through the ankle's bone protrusion. Outward seams are determined by measuring from the natural measurement of the waist to the floor on the opposite side of the body as shown in Fig. 5.10.

Once you've obtained the necessary measurements, you'll develop your designs depending on the whole design and fit intent. How will you do with these measurements at all?

How Might 3D Measure Up's Standard Body Measures Be Utilized?

A 3D Measure Up will collect more than two hundred evaluations, including the body's diameter, width, length, and dimension, in addition to the placement of significant areas like the waistline, thighs, and the shoulders. The information gathered can be utilized for creating an electronic representation of the body, which may be utilized later for creating custom-fitted clothing. Furthermore, assess movement and posture, and monitor modifications to shape of the body as time passes. In summary,

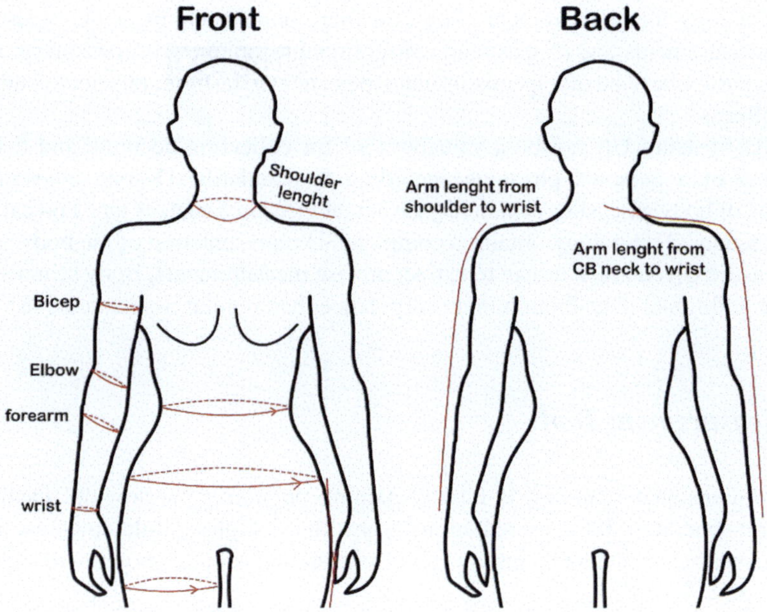

Fig. 5.9 Elbow length, bicep diameter, and shoulder length measurements' points

Fig. 5.10 Knee length, ankle diameter, and thigh measurements' points

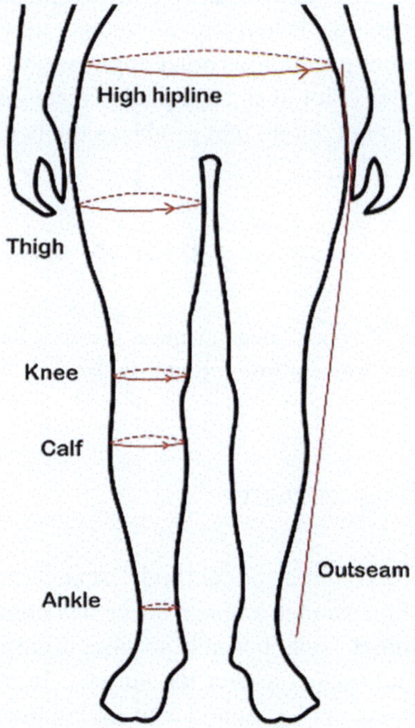

3D technology for measurement allows for more precise and highly computerized body measurements and shapes than conventional techniques for measuring, and it may be particularly advantageous in industries such as fashion, physical condition, and wellness.

A 3D Measure Up can be a valuable tool for collecting accurate and detailed measures of a person's physique according to standards. Despite conventional methods of body assessing, which rely on actual measures such as tape and calipers, the 3D scanner technology obtains a complete 3D representation of the body, which 3D Measure Up can then utilize to extract precise measurements. Body scanners can provide additional data about human corpulence than regular assessments [8].

3D Measurement Tool

Three-dimensional scanning is rapidly gaining popularity worldwide! Dedicated software generates 3D representations from photographs, while low-cost three-dimensional scanners enable fast and precise scanning. Mobile phones will someday contain a 3D scanner.

Three-dimensional body scanning is the most effective tool for recording a complete body or additional components in 3D.

A 3D body scan is a three-dimensional representation of the human body or limb that is projected onto a screen. It may be employed to examine one's body and stance, among numerous other applications. The three-dimensional representation may have been colored and sculpted to resemble the original, or it can simply be a mesh, a form of point cloud that produces the body.

Advantages of 3D Body Scanners

A 3D body measurement scanner has many advantages over standard body imaging and measurement technology.

Non-intrusive

Three-dimensional body scanner requires no intrusive procedures. The three-dimensional scanner never has contact with the subject being scanned. If a three-dimensional portable scanner is employed, the person who operates it sets the device and rotates it over the subject. In a three-dimensional scanner, a stand the subject being photographed is placed in the middle of the room, and a variety of sensors put in the enclosures may capture each characteristic of the subject in three dimensions, from every vantage point.

The Most Rapid Scanner

A three-dimensional scanning machine will identify an individual in approximately 10 s. The operation will only require a few minutes with a portable three-dimensional scanner, and the individual in question must be completely immobilized.

Being accurate, three-dimensional scanners are meant for taking accurate and automated measurements of the body. Three-dimensional scanners for the body are extreme-precision devices that produce detailed models in three dimensions at extremely high resolution. They can't identify things that the human eye does, yet in numerous scenarios, three-dimensional scanners are quite beneficial. Obviously, the quality of the results depends on the body scanner. Some three-dimensional scan stands are capable of gathering many different individuals simultaneously, including colors and textures, with a staggering quantity of precision.

Types of Three-Dimensional Body Scanners

Three-Dimensional Scanning Booths

Three-dimensional scanning booths are sensor-equipped cubicles. The individual positioned in the focal point maintains the position for a few moments in order to the body scanner can capture pictures from every perspective simultaneously. A three-dimensional figural body scanner delivers an intricate 3D representation of the human, encompassing texture and color [9].

Handheld 3D Scanners

Portable three-dimensional scanners are considerably more versatile than three-dimensional scanning booths considering that they could be positioned around the user to photograph their whole body. The drawback involves the fact that the individual being scanned in 3D needs to stay still entire the entire processing execution; alternatively, the three-dimensional scanner would be ineffectual. Portable three-dimensional scanners are frequently utilized for capturing body portions or appendages in body scanning 3D applications, as opposed to full three-dimensional body scanners [10].

The Applications Involving Three-dimensional Scanners

Three-Dimensional Miniatures: These minuscule actual situations sculptures, additionally referred to as three-dimensional selfies or three-dimensional pictures, frequently are a three-dimensional created utilizing a full-body 3D scanner framework, which is generally a three-dimensional scanning booth.

Fashion/Apparel: The apparel industry utilizes three-dimensional body scanners to design customized and adapted clothing. Alongside the introduction of virtual fitting centers in retail, 3D scanning of the body opens up intriguing possibilities for both in-store and online shopping: Customers can take advantage of their three-dimensional avatar, that has been built on precise measurements of their bodies, to immediately attempt on fresh clothes.

Monitoring Appropriateness and Body Outline: Three-dimensional have been employed in sports for monitoring the growth of an individual utilizing numerous body regulations, contouring statistics, postural assessment, and others.

Since 3D scanning results differ from manual measurements, how can we convert them into garment or pattern-making sizes?

The most intuitive way is to collect big measurement samples and then convert the results of 3D body scanning to manual measurement.

As a result, we measured the busts of 24 men and 23 women using a 3D body scanner and a tailor's tape.

We evaluate two sets of data and derive a formula ($y = 0.1183x - 7.8163$) for converting 3D scanning findings to manual measurement results as shown in Fig. 5.11.

The formula allows you to simply convert three-dimensional body measurements into manual measurement data, which you may subsequently use to convert into garment or pattern-creating sizes.

However, the table is mainly for reference. Every tailor has various measurement habits, which result in different conversion formulas.

Determine a Transition Point Between 3D Body Scanning and Garment Sizing

However, turning 3D scanning measurements into manual measurement data is not the primary goal of using a 3D body scanner. This is because, even with manual measurement data, numerous sessions of fitting and adjustment are required, which takes time for both the tailor and the customers.

Our primary goal is to identify a transition point between 3D body scanning and garment sizing, give solutions, and introduce fresh innovation into the apparel sector. Take our client "Facha" in Taiwan as an example.

Facha is one of our top clients, successfully utilizing 3D scanning technology to increase product value and service quality. Due to time constraints, a tailor could previously only collect approximately 15 sets of body measurements. A 3D body scanner could provide the tailor with over 200 sets of body measurements. How might they use this information to improve their service and give ideally fitting clothing to their customers? To answer this question, Facha began a two-month experimental program. During the program, they ask all of their customers to complete a 3D body scan in addition to physical measurements taken by the tailor.

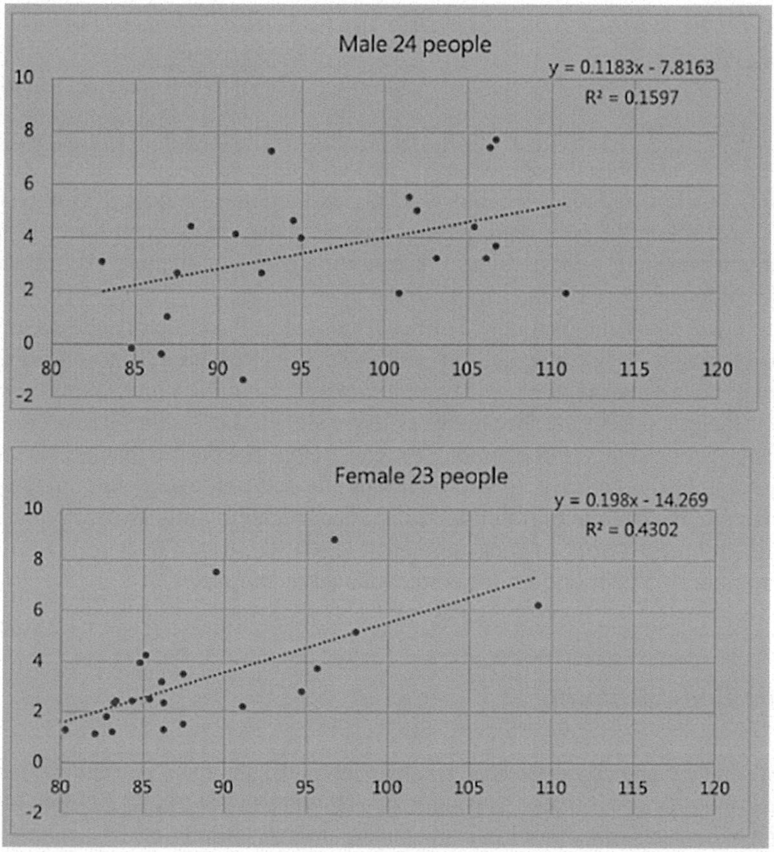

Fig. 5.11 Graph for the 24 male and 23 females for formula derivation

They collect vast volumes of data, convert the scanning data into pattern-making data, and lower their garment alteration rate from 40% to less than 5%.

They successfully make the most of the 3D body scanner to improve client experience while also ushering in a new chapter of the fashion future through technological advancements.

Three-Dimensional Body Scanner and Manual Measurement Comparison

In regard to reliability along with precision, three-dimensional scanners surpassed traditional calculating anthropometry. Particularly, the interobserver variances (accuracy) for waistline and hip diameters were higher than those acquired with the three-dimensional scanner (traditional anthropometry: waist: 3.9 cm and hip: 2.4 cm vs. 3D scanner: waist: 1.3 cm and hip: 0.8 cm). Apart for hip diameter, no substantial variance among repeated three-dimensional assessments.

Automated assessments found significant variances. Additionally, the accuracy of the measurements taken manually exceeded 2.50 cm, reaching 8.19 cm, indicating an even greater disparity.

Nevertheless, additional investigations have shown that scanning procedures do not enhance consistency. For example, Bragança et al. discovered that the 3D scanner was less reliable and more precise than traditional anthropometry, with greater standard errors of measurement for all metrics except neck circumference. Furthermore, both the technical errors of measurement (TEM) and relative technical errors of measurement (%TEM) outperformed the manual technique [11].

Mobile Body Scanning

The most recent advances powered by artificial intelligence science and technology, such as computer systems and deep algorithms for machine learning, have enabled the creation of contemporary mobile scanner technologies. Usually, body scanning structures demand the client to take multiple photographs of themselves before following their personal information to an organization in a matter of seconds or minutes. These technologies bring substantial advantages to customized firms and their clients, such as clarity, mobility, and expenditure reduction (Fig. 5.12).

3DLOOK's mobile custom online assistance solution permits customized enterprises to obtain accurate body measurements from their customers virtually. The device, which requires no technical understanding to be operated, creates points of measurement in under a second from just two photos taken by the user. Everyone in the company is able to utilize the measurement data through a customized workspace. Businesses that employ three-dimensional fashion design software can also receive their customers' customized dimensions of body representation.

One of the advances that mobile scanning technology offers to present-day tailored enterprises is the capacity to instantly offer supplementary significant information about the consumer that goes outside measurement. As an illustration, three-dimensional models and body form data can be separated by location, size, age, race, and gender [12].

Customers may practically assess themselves anywhere using mobile scanner technologies that enable enterprises to cater to their requirements while conserving

Fig. 5.12 3D VLOOKS

both time and resources through accurate measurement. Additionally, smartphone scanners are capable of measuring undressed folks, enhancing ease and comfort.

Conclusion

Recognized fashion the company's designs must be distinctive and innovative while maintaining an exceptionally high standard of quality. Appropriate body measurements are essential for appropriate clothing production. This maintains the cloth while simultaneously guaranteeing that the collection has an appropriate body form.

Inadequate body measurements diminish the material of fabric and, in certain circumstances, damage a valuable collection. However, fixing inaccurate sizes takes a bit more time. It affects labor, resources, and, eventually, the fashion brand's image. Precisely an outcome, every collection that becomes accessible has to incorporate appropriate sizes. Three-dimensional scanners appear to be a valuable technology for body assessments, as they provide computerized, quick, and simple assessments of circumferences, body volumes, and fat mass, among other features.

This article shows that body scanning could help guide pattern design processes. However, when creating garment patterns, it is essential to account for the application of measurement and create landmarking accordingly. Three-dimensional scanners appear to be beneficial technology for body evaluations, as they provide computerized, quick, and simple assessments of diameters, and body dimensions, among other characteristics.

Three-dimensional scanners seem to offer repeatable, precise, and secure findings which are consistent with the various methods utilized to evaluate body measurements. Future research should focus on studying both people with body composition concerns and those with an elevated mass index. Landmarking factors comprise the points that mark measurement areas. There's an important difference among traditional techniques which employ the body's cartography (which is frequently referring skeleton landmarks) and body inspection, which is based on the surface's geometry and must determine landmarks exclusively from it. Although specific studies preceded landmarks (such as CAESAR), scanning one's body is prohibitively expensive and requires self-control, but is exceedingly time-consuming. Improvements in landmarking criteria might eliminate a few of the possibility of inaccuracies, enabling manual and scanner procedures to identify landmarks and evaluate comparably.

References

1. B.E. Jones, Measurement: past, present, and future: part 1, in *Measurement History and Fundamentals, Measurement, and Control*, May 2023
2. S. Gill, A review of research and innovation in garment sizing, prototyping, and fitting. Text. Prog.Prog. **47**, 1–85 (2015)
3. S. Gill, M. Ahmed, C.J. Parker, S.G. Hayes, Not all body scanning measurements are valid: perspectives from pattern practice, in *The 8th International Conference and Exhibition on 3D Body Scanning and Processing Technologies, Canada*, October 12 (2017)
4. E.C. McKinney, E. Bye, K. LaBat, Building patternmaking theory: a case study of published patternmaking practices for pants. Int. J. Fash. Des. Technol. Educ. **5**(March 2015)
5. M. Ahmed, T. AlrushaydanL, S. Gill, S.G. Hayes, K. Brubacher, *The Suitability of Body Scanning Measurement in Pattern Drafting Methods*, Switzerland, October 22–23, 2019
6. Standard Body Measurements Mar 6, 2023, online source
7. S. Thiel, What are body measurements and how do you figure them out? April 18, 2020
8. E. Bye, K.L. Labat, M.R. Delong, Analysis of body measurement systems for apparel. Cloth. Text. Res. J. **24**(2), 66–79 (2006)
9. S. Gill, A review of research and innovation in garment sizing, prototyping, and fitting. Text. Progr. **47**(1) (2015)
10. Training Institute Pune, Overview of 3D Body Measurement Scanners, Feb 9, 2023, online source
11. L. Rumbo-Rodríguez, M. Sánchez-SanSegundo, R. Ferrer-Cascales, N. García-D'Urso, J.A. Hurtado-Sánchez, A. Zaragoza-Martí, Comparison of body scanner and manual anthropometric measurements of body shape, June 8 (2021)
12. The evolution of body measuring, How technological advances are driving the future of made-to-measure fashion, Online source

Chapter 6
Analysis of Bodice Block

Muhammad Saqib

Introduction

The topicality of the bodice block concept (common practice for the fashion industry as the "basic bodice") in the field of garment construction generally and English pattern making in particular is significant and is one of the most important consideration by a left conception. This basic pattern will be a central part of the creation process for an assortment of garment styles, starting from the most basic top line shifting toward expricancy, the characteristic feature of couture fashion. This chapter focuses on all the complicated and refined issues related to tracer, tracing, and sewing of the bodice block. This means that it attempts to make sure that readers gain the full understanding of those aspects which make it essential in both tailoring and the full context of the garment design [1].

"Sewing" is the "dictionary and imagination" of a patternmaker, his or her initial outline on a piece of fabric, a basic form from which the "texture and volume" of fashion are molded. The painstaking procedure of making bodice blocks could be described as both an art and science that requires you to have a top expertise in accuracy, detail, and anatomical correctness. Use our AI to write for you about any topic! Such basic principle is not just a template or a platform, but a tool with which a designer realizes one's ideas into clothing pieces that are then transformed into the tangible embodiments of art and aesthetics.

Educating beginner on how to master bodice blocks starts with understanding the essence of each component—they are all front, back, and sleeve blocks—which are equally as important together as they are individually for the dress to have its unique fitting and appearance. A basic bodice block made by hand holds a special place in the training of every aspiring pattern maker novel here is a process that requires detailed calculation of one's measurements to achieve a flat pattern that accurately

M. Saqib (✉)
School of Arts and Design, National Textile University Faisalabad, Faisalabad, Pakistan
e-mail: saqibrao@ntu.edu.pk

M. Q. Khan et al. (eds.), *Garment Sizing and Pattern Making*, SDGs and Textiles,
https://doi.org/10.1007/978-981-97-7683-2_6

reflects the real proportions. This section will take readers through the intricate steps of designing a pattern for their apex quilt bottom, reiterating the significance of precision in taking measurements, calculating darts and allowances, and directly coming up with a pattern that clearly represents the designer's idea.

Worth knowing how to make the bust fit tightly is the very first step toward success. This chapter goes in detail on the science of fitting, paying attention to key areas of fit like the bust, waistline, and seam lines, and how it is possible to tweak the exact fit by adjusting these parts. A sagging bodice is the most visible "hallmark" of a low quality tailoring; resulting in a number of discomforts, lack of mobility and aesthetics. Readers will get a grasp of fitting issues and the ways in which they can then address them, thus making their tendency to end up with spotless finished garments that not only look beautiful but feel right as well, more consistent.

The bodice block is one of those things that possess particular versatility, and so it is one of its most prominent features. In this segment of this course we shall demonstrate that an elemental basic basis of pattern can be developed into limitless design concepts. Instruction by instruction readers will learn how to work on the block of the bodice so as to create a good fit whatever shape one wants and to add details such as necklines, yokes, or fullness any time needed. Here, we strive to trigger the imagination and spark innovative ideas, empowering readers to explore the broad area of the available options.

Through the development but not only at the stage of drafting a pattern, the manufacturing of clothing is a very complex process. Fabric selection, preparation, and the relationship between the fabric and the pattern are imperative elements in the design process pool between the idea and the finished product. This chapter will focus on these aspects, giving the reader information on selecting the best fabric for different designs and also testing/preparing it according to the pattern and the design.

Concepts behind the elaboration of the details such as adding pleats, tucks, and princess seams are also discussed intensively in this course. Today, they are less prevalent than traditional models, but they do offer a pattern maker an opportunity to impart additional items in the aspect of fit, function, and aestetics into the product. Practical and real-word applications with some stitched fabrics and case studies will further demonstrate the bodice block ability to be versatile, citing examples from modern fashion that can with the interviews from industry professionals.

To conclude, the last part of the chapter will probe the path of bodice block design evolution and its role in the midst of the fast changing and transforming spheres of fashion. Innovations in technology, shifts toward environmentally friendly practices, and emerging trends in designing are going to affect the principles of using and making patterns, hence, when the pattern makers, both new and old, are made to deal with new challenges and housing these challenges in the old ones.

The bodice block is not just a mere pattern that is being borne in mind, as it is the very floor slab on which the construction of the whole pattern is based. This chapter is devoted to guiding the apprentice through the examination of a bodice block's construction, fitting, and adaptation. At the end of it, the learner will be left with a technique tailored to their own bodice block design. This method will open doors for tailoring of garments as well as improved pattern making in the near future. By getting into the details of the bodice block we trace our way through the process of creation where exactness and imagination speak to each other and the very art of customizing evolves and strikes us something serious.

In the garment industry, specifically in countries where craftsmanship is embraced, where English seamstress pattern making rules the roost, the creation of a women's bodice block reflects the art of meticulousness. The connection of these simple pattern pieces is available through a number of techniques and forms. In this case, the jacket comes without the collar or embellishments such decorative sleeves. This simple design is the foundation which any item is built on to achieve perfect wearers fit and realization of the designer's vision. The road that starts from a concept being gone through the actual measurement and artist interpretation is all connected with each other, it is rally as important as each step of the outcome.

The Essence of the Bodice Block

At the center of the bodice block, the makers were given another neutral picture so they could start decorating all to start creating styles using high and low neck necklines, various widths for sleeves, and all kinds of possible neckline. It is described as a constructed, skeletal frame which the artist precisely put together from the neck to waist to create individual sized patterns for making the garments. The importance of the footwear only can't be underestimated, it is the basis on which the criteria of fashion, style, and fit are build upon [2].

The precise art of a bodice block includes knowing the anatomical shapes, the ability to discern good form and silhouette, and a hand that produces a straight and uniform block with precision. The goal is to transform into a simple contour in two-dimensional shape that, while on wearer's form, the strange irony is that it would have to be a perfect fit for the three-dimensional figure shape in order for wearer to be comfortable.

Measurement and Drafting

The initial stage of this precise operation includes the process of measuring the upper half of the body by all measurements. Accuracy is instrumental here, because the outcomes of precision is the figure on the final garment. It is not in depicting flat and surface edges, every lines, and curves of the body; it is the process of rendering them into paper as volumetric objects.

Perimeter (girth), in addition to critical lengths, which include shoulder to waist and the back neck to waist, are also among the variables taken into consideration. Once the front and sides have been established, the blocks allow the designer to cross the intellectual/accuracy gap and get to the stage of creating a wearable pattern.

Crafting the Bodice Blocks

This development of 2D patterns into actual shapes begins with drawing patterns with pattern paper on front blocks and back blocks. This phase is a very subtle process that involves a lot of precision in geometry and the understanding of the body shape and curvature of clients. This is meticulously achieved by carefully picking and connecting points as determined by the collected measurements, implementing a careful mix of straight and curved lines and other mechanisms, such as darts.

Front and Back Blocks: A curvy silhouette for the chest is achieved with the front panel that has darts for the bust, and it reaches to the waistband level. The supporting garment that's specifically designed to adapt to the back's natural contours comes in a form of as the "back block" which has the modifications that accommodate the flatter basic structure of the back as compared to the bust [3].

Sleeve Block Integration: Another saved for later the sleeve part of the bodice block, the sleeve block, is that the best fit into the armhole is ensured and it is better to transfer the block to another pattern flat. Here, you have to be very smart in order to get exactly, how the arm moves and, how effortlessly it affects the form fitting and the general comfortability of the garment.

Refinement and Testing

After this, fabricating a toile, a copy produced from an inexpensive covering is the next essential phase. This model helps the pattern maker to test nodes of the bodice block on the real body, and a body-length view of how the pattern comes out from paper to the person. It is exactly here, at the point where alteration, finishing, and making perfect happen, that real pattern making wizardry starts. Whereas fittings will provide hints exactly about how tight or loose the garment is, as well as ensure that the darts are properly aligned with the natural contour of the body. Each observation, meanwhile, is a puzzle to be solved, and the pattern on the toilets is tuned by moving the designs from the toile to the models and vice versa.

Final Adjustments and Pattern Finalization

The final stage where the bodice fitting advances and the bodice block is made into its final production sample is what this stage is about. Having learned everything from fitting the toilet, it is widely a thing that we feed the pattern with to ensure that it is not only fitting. Rather, it is a kind of a base for the following designs. It is the final savior block that I had trusted as the template block from which thousands of varieties of the garment exteriors can be created with the full assurance of both consistent and flattering fit.

Apart from its functional aspect, the making of a block is a very deep learning journey for any patternmaker who tries to get through the long and detailed times of finessing and redoing. It sharpens their skills of measurement and drafting, broadens their concepts about the human anatomy, and builds upon the development of a visualization and an ability to make complicated designs. This too reflects the complex relationship between design and the best practices of the art of tailoring, where technology and creativity come together to give birth to such patterns that make outfits beautiful and comfortable to wear.

Toward this regard, bodice block acts as a blueprint for fashion designers and tailors as it illustrates the detailed peaces that is put together into every garment. It sends a message about what one values by going comfortable and fashionable, the work they've put into it, and showcasing one's love for tailoring. By means of these rigorous stages pattern makers lay down the basis for the two requirements a pattern must meet: it not only has to look nice but also fits well. A trip from simple measurements up to a detailed bodice block is indeed a representation of the era garment construction, a combination of the techniques of past ages but together with the modern design goals [4].

Constructing Basic Bodice Block

The first step in the garment construction process, tailoring a basic bodice block from the ground up, will be the most significant point in onwards development. This blending of precise measurements and the artistic skill of patterning is the core essence of the garment-making world. This, in particular, is the key to any product's quality which must meet the criteria of both perfection in size and style that is tailored to the client shown in Fig. 6.1.

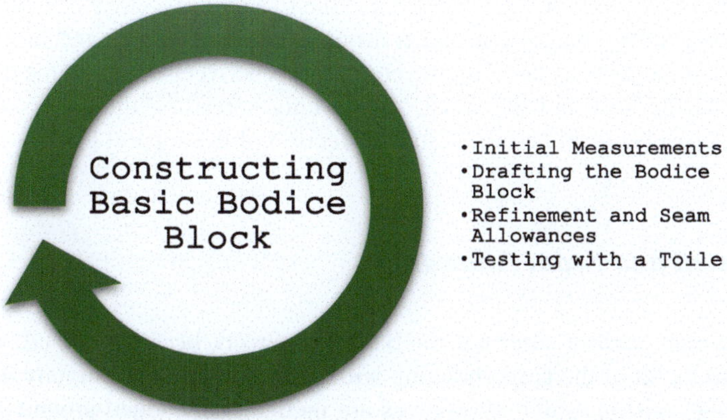

Fig. 6.1 Steps for constructing basic bodice block

Initial Measurements

The approach first requires identifying the correct points on the body, a process which implicitly ensures the entire create procedure stays on track. This job is all about being accurate because if the tiniest deviation from the size will not match properly. Entitled with a retractable tape measure, the designer attempts to record the chest, waist, and the hip circumferences and these measurements are made without inculcation as well as no use of any constrictions on the body.

Besides just these values, the horizontal distances like shoulder to waist, own waist, back neck to waist, etc. are the most important ones. They give a precise clue about the length of our torso which is an important consideration since it determines how the shirt falls on the body. An arm length together with the width of shoulder and neck have to be measured and they all are giving a hint about the length and width of the body, accordingly when the garment is being molded to fit the shape of the wearer.

Drafting the Bodice Block

By which means with quantities in our head, we move to the next step during that converting measures onto pattern paper imprint with the marking of the beginning of the bodice block's physical realization. This step is where artistry in pattern drafting joins the detail of anthropometric measurements. The designer guides the pattern maker by plotting the points on the paper based on the measurements taken. For the straight up/down lines, which stand for the center front and back, the designer uses straight lines and for the armholes, neckline points, the designer uses rounded lines making sure that the pattern follows the body contour.

At this point, darts happen to be the main tool used for altering the dimensions of the pattern, so that the finished garment is able to serve as a close fit with the curvy shape of the body. The pattern maker would determine where and how to locate these darts in the pattern based on the comparison between different measurements, for example the bust and the waist, thus evenly spreading the excess fabric that contributes to a better fit and the visual perfection of the apparel [5].

Refinement and Seam Allowances

The refinement process, after has the basic framework been developed, will adjust the draft so the fit of the garment along with the aesthetics is accurately depicted as the designer's vision. Stitch allowances are meticulously added around each piece of the pattern. This achieves a space that can be sewn together properly to close the garment without distorting the fit. This part is a little bit meticulous in which you are going to draw marketers carefully, because incorrectly marked allowances may lead to a finished clothes that is too tight or too loose.

The neckline and armholes are those spots which take a considerable amount of time to finish. Wearing of trousers, which are ironed and adjusted for comfort flat laying without either pulling or gaping. The different shape and depth area of the neckline can level up or even eliminate the elegance depending on the design, while the selection of the cut of the armholes needs to fit the degree of comfort without large amount of the material bunching.

Testing with a Toile

So, what is the last part in the process of custom-made bodice making? It is the stuffing and fittings of a toile that is a prototype of a toile clad in inexpensive textile. This sloper is used by the designer in order to examine the pattern's fit not only on a paper draft but also on an actual body, enabling the designer to observe the clothes' movement. The sloper is designed to serve as a precisely fitted dummy and for that reason it is a basic area draft that still has lots of room for excess. This phase is essential for the designer because he can spot the points where the garment accommodates too much wrinkle, too loose or what will not line on the body's natural lines because of the darts and seams athletically accommodating.

Pattern Fitting—There may come a time when the alterations leading to the curves of the neckline or the armholes or some minor adjustments to the darts or the seam allowances are required. Every change from the base bodice block is precisely transferred back to the final pattern, which in turn, ensures that the final over all fit and look to be just like the intended style [2].

In like manner, this process showing detail and nuance which is made up of the four phases will be its capacitor tool and at the same time, it will be an alley to its multitude creations. This process, in particular, sharpens the designer's skills in measuring and sketching but also leads to a holistic comprehensions regarding the attributes of a fabric and how it drapes around the human skin, putting together technical understanding and creativity, to make a dress that is as admirable as it is comfortable to wear. The production of a bodice block of the simplest form, however not so straightforward in practice, truly represents the essence of garment-making, as it incorporates traditional techniques with the modern taste of aesthetics.

Anatomy of Well-Fitted Garment

The anatomy of an equally fitted bodice lies in the middle of the key principles of garment construction which is the simultaneous consideration of precise measurement, apt pattern drafting, and deep conviction in the human anatomy. In the sphere of the fashion design, the art and science of the correct fit is and will be a challenge, and the hard points on a bust point or a waistline, and an armhole is taken in the process. A lovely dance between the fabric and shape is a process that can't help but being subject to our objective assessment of what qualifies as a proper fit and where adjustments to these key areas can make a huge difference [6].

The Foundation of Fit

Measurements demonstrate the form of a core of a well-cut bodice. Each circumference: Bust, waist, and the torso length, is a foundational beam that holds the whole body. The bust point, also referred to as the apex, being the main spot of a perfect fit, it must be perfectly placed on the garment so that it accurately covers the chest and visually pleases the eye. The bust point misplacement in the pattern makes it possible that the fabric across the chest where the bust sits may either be pulled tight or being loose that is hanging improperly, forming an unsuitable profile. In this context, the waistline, which equals the narrowest point of the bust, is also a crucial measurement for the garment to fit well at the tightest part of the torso since the shape of the body transitions smoothly at the waistline without disrupting body movements in any way.

Sculpting the Bodice

The subsequent step in this pilot is collection and analysis of the basic dimensions which will in turn result in the drafting stage where your surface will come alive from its flat plane. The detail of sleeves sews in this stage while the holder or armscye becomes its central point. The armhole curve is highly technical part of bodice construction where you must ensure that the fabric falls gracefully and you have ellipsoid seamless lines. The drafting process further consists of pinning and shaping of darts which are used in order the tailor the cloth to the bust and waist area leaving just the right amount of space and thus an exact fit. The darts are the most crucial esthetic element of tuxedo jackets. These darts go, lengths, and positions are critical; they must be calculated to channel fabrics toward the bust point while tying off elegantly toward the waist, thereby achieve a look that flatters without squeezing.

The Art of Adjustment

The ultimate test of the bodice fit, nevertheless, lies in the moment when all the drafts and measurement techniques that have been theoretically employed finally meet the real live body, and the order forms the unique outline of the human anatomy. A careful examination of the budget will narrow down the critical features that fit at this phase. An entity of patterns should have the armhole reshaped to ensure there are no discomfort or fabric bunching. Doing this would help the sleeve to join smoothly to bring out the intended silhouette of the bodice. Modifications may be made to the bust area and at the waist to correct any fashion issues that could result from the proportion discrepancies between the pattern laid out and the actual dimensions of the wearer. It may be changing the level of departure or pointing the arrows slightly, or maybe even repositioning them taking into account the natural shape of a mind. The repeated turnaround of taking, adjusting, and readjusting consistently appears to be expedient as it substantially helps in the finalization of the bodice into its most astounding and comfortable style.

Harmonizing Form and Function

What follows these carefully put adjustments is breathtaking bodice which is not just a perfect fit but also beautiful whether you wear it or a person's wear it. But to add ease and look wonderful on the wearer we designed the bodice to add comfort and at the same time ease of movement. A bodice that is cut accurately sets the tone to ensure the sleeves and skirts or trousers are free-flowing, non-fitted and balance the garment's design. It is a never-fade demonstration of the designer's functionality-styling effort aimed at making sure the fabric and the shoes look good as well as feel good. The bodice's petitpas stages the process of the fit, beginning with the exactness of initial measurements underscoring the requirement of precision in creating fabrics that do not appear bloated. This process involves attention to detail and the final tinkerings accentuate this. Here, it illustrates the significance of every step starting from measurement capturing to final stitching beyond mere seams. However, it is the symphony of them having created a garment fitting the highest standards of fit and fashion.

Tailoring a beautiful bodice is a task, both tricking and tricking for a fashion designer, but it is also a great occasion of a designer's skills. Certainly, it involves not only technicalism as well as exactitude, but also an intuitive feeling of how the fabric behaves in the human form. This is a balance via measurements, adjustments, and drafting with sewing so designers are trying to have anyway that makes him/her look spectacular and feel like a second skin to the wearer. Through this dynamic, the drawn-line of fashion design from start to finish is evident and this art and science reveals the high level at which fashion designers work, jealously guarding the level of excellence.

Adapting the Bodice for Various Designs

Adapting a basic bodice block to various designs is a nuanced process, blending art with technical skill to craft garments that range from the elegantly simple to the complexly artistic. Let's delve deeper into each adaptation technique, providing richer examples to illustrate the transformative power of these methods.

Dart Manipulation for Design Variability

Manipulating a dart is a fundamental element in modifying the shape and offering these new elements.

Creating Princess Seams: You could, perhaps, take a traditional bodice and detach it from the rest of the dress so that it will have princess seams. Work your way through identifying the locations of darts on your basic bust block. The pleasure of creating an exquisite garment starts with the "slash and spread" technique. The dart's entry is secured at the point, and the bust and waist cut is parallel to the pattern admission, so that the seam curvature at the bust gently falls to the hem. The absence of the waist dart, and instead, the self-sculpting of the fabric around a shaped body, achieves an equally fitted result whether for casual or formal dresses.

Introducing Gathers at the Neckline: For a romantic feel, you can rework the bust darts in a blouse, so they add pleats at the neckline. The dart direction should be altered toward the back neck. This will remove the dart and produce more edge fabric that can be gathered easily. That way the initial collar design becomes product's feature that, although it may seem simple at first glance, actually has its volume and softness about it that beautifully frames the face [7].

Adding Fullness and Flares

Reaching this goal can be achieved by using volumetric pieces low by fitting or flaring up the item, which will facilitate good motions and volume.

Creating a Peplum Blouse: In order to change a bodice block into a peplum blouse, upgrade and slash the bodice block until you reach the needed peplum length and spread it from the waist down. The voluminous effect makes a peplum which can be pleated or gathered, a wonderful detail to level up the outfit with elegant playfulness (Fig. 6.2).

Designing a Flared Dress: Convert a conventional bodice into a flared dress by extending further and then slicing upward from the waist downwards until you get to the hem and spreading out automatically. It can be used with an asymmetry effect, where the flare starts on a higher level on one side, yet looks great in the way that you combine a high-low effect that is a fashion icon you want to be remembered for.

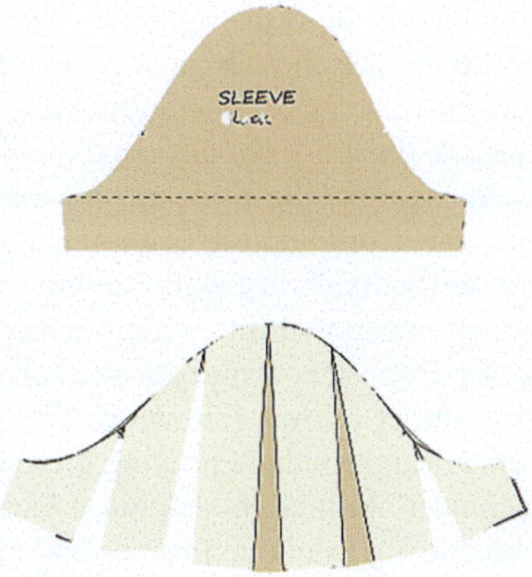

Fig. 6.2 Adding fullness and flares

Incorporating Design Elements with Seam Lines

In a strategic way, seamlines can integrate own structural and decorative elements to the garment.

Asymmetrical Bodice Design: Visualize the bodice block constituting the top with a cut in a crossover style, starting from one shoulder and ending at the opposite point of the waist. The creation of a waistband can also be made by securing darts into a hoop that spans the waist, introducing a moveable art form that travels the fiber taking the subject from one element to the next and offering a space for change of colors or texture.

Empire Waist Adaptation: Stitch the waist seamline slightly higher up to the bust and distribute the intake of the dart from this higher line to do empires waisted tops or skirts. It is definitely an asset for me that I was to get used to this adjustment, as it enormously helped me be more comfy while being at the same time attractive.

Experimenting with Sleeves and Necklines

Highlighting the significance of sleeves and necklines in a piece by piece design, we can definitely deduce that they are the elements which create and maintain a specific look and feel it has.

Bell Sleeve Addition: For bell sleeves, take your original sleeve block and slash both ends from the hem to the cap, then spreading the patterns to create fullness for the sleeve that will peak at your wrist rather than end at the cuff. Having the ability to do crossover collars can easily transform the very basic long-sleeved top into a must-have outfit for those who love a bohemian or the vintage-inspired style.

Adjusting for a Boat Neckline: Instead of the traditional V-shaped neckline, make a simple change by drawing a new horizontal straight line starting from the left shoulder as far as the right one. Thus, it will broaden the neck area for an open boat neckline. The modification demands that this portion of the shoulder be redrawn carefully and possibly slightly increased in terms of the size of the armhole to make sure the dresses have the same fit with or without the V-fold neckline.

Speaking of the artworks, you come to understand adequately how a bodice block can start from humble beginnings to be transformed into something elaborate and extraordinary. Every skill ranging from designer manipulation of darts to the invention of new possibilities such as sleeves and necklines to garments suit the different tastes and offers a myriad of choices for designers to come up with the best and also attract a loyal following. The process of translating the block for the bodice is the design studio's proving ground, highlighting the designer's inventiveness and technical abilities in shaping a variety of garments, as well as the proficiency of transferring from one form to the other.

Common Fitting Issues and Solutions:

Creating a well-fitted bodice is a cornerstone of garment construction, yet it is fraught with challenges. The path from draft to a perfectly fitting garment is often a complex process of addressing and correcting common fitting issues. This detailed exploration delves into these challenges, offering solutions through a combination of technical adjustments and creative problem-solving.

Gaping Necklines: Identifying Causes and Crafting Solutions

To begin, bodices may have a too-deep neckline that makes them look disarrayed. This is a serious matter in the overall perception of a garment's appearance and style. Quite often, the problem of such a misfit is due to an error between the curve of the curve and user's body, particularly in cases when the bust and shoulder measurements are precisely not proportional to the pattern [8].

For the case of a round neckline where a slight gaping is always a usual situation, may it be fixed by only taking a little bit the inside of the seam to raise a neckline and bordering on its circumference.

The V-neck detail can be enhanced or refined with small darts or tucks. They assist in molding the body. Likewise, these changes would conflict the original idea of neckline; hence, more secret adjustment like altering the dart angle or depth will be more suitable as that will demonstrate the way the fabric is laying against the body more eventfully.

Tight Armholes: Adjusting for Comfort and Movement

At the same time in addition the tops which have the minimum armholes not only show off as a discrete inconvenience but actually lead to the restriction of movements which finally make the clothes inappropriate for wear. The fit issue is more over often the results of no ease in the bodice or an out of alignment sleeve head.

Increasing Armhole Depth: Adding more depth to the look for the armhole in the base pattern of the bodice will make the arm feel better. This process includes adapting the sleeve pattern to the new armhole contour; as well ensuring enough ease to avoid a tight sleeve.

Adjusting Sleeve Cap Ease: Could removal ease from the sleeve cap help for relieved scratch around armhole. This is having the seam of the sleeve cap buried to reduce its thickness and make the sleeve to set into the armhole without affecting the bodice shape.

Bust Dart Misalignment

Unsuitable bust dart alignment can result in the upper panel of the bodice that either disturbs or flattens the bust area. In bodies sewing the position of pins is important to ensure that the cut is the right one and look good on your figure.

Repositioning Darts: To move the darts which point up or down of point, remeasuring the bust point and adjusting the dart's end point could help in making their dart apices align with the wearer's bust apices. This tugging of the elastic edges allows everything to distribute evenly across the bust, to provide the most smooth look.

Splitting or Rotating Darts: On such occasions, instead of a single arrow, two or more might be needed for shaping. If the back fitting is too loose dividing the flat pattern in two pieces makes the dart shorter and thus more fitting. An alternative way is to rotate the dart intake moving it toward the waistline or side seam.

Addressing Back Waist Gaping: Ensuring a Smooth Silhouette

Back cumbersome and wide gaping on bodices is general, and could ruin the garment's shape, especially in designs that require a tight fit. Characteristically, the problem develops due to different kinds of bodices curves from those of current body shape.

Adding Back Darts: Putting darts at back waist can be considered an effective way to supplement bodice contouring with the figure accordingly, making the gap disappear. These will vary holding the bra on the chest in this way will allow to reevaluate their depth and placement on the basis of individual body shape for best fit.

Adjusting the Side Seams: Taking in the side seams at the waist and as well may help you with gaping, though this might not be a very good idea for fit of the bodice. In-depth modification is indispensable for the perfect fit of the garment and the possibility to avoid unpleasant extra restrictions.

Side Seam Imbalances: Correcting Fit Across the Body

While the unmatched side seams, where one side of the waistline fits differently when measure than the other, are the most difficult to solve. This may become an issue since the piece of clothing might not fit a person well due to the pattern inaccuracies or the asymmetrical body shape of the wearer.

Individual Fitting Adjustments: The humanization of the bodice can be done if the sides are custom fitted to cinch into the wearer's body. This helps with any side seam imbalances. This may be done by getting in at or letting out the side seam of the bodice on the opposite side, which ensures that each side is fitted just for the particular shape of the wearer.

Asymmetrical Dart Placement: This is especially helpful for bodies that have considerable asymmetry, as it allows to apply darts different on each side of the bodice in order to compensate for the arms at fault. It is about using a ruler and other tools to make the alterations while ensuring the changes are not weird and functional.

Overall, frequent fitting problems like a big gap behind the neckline or tight armholes, the bust dart misalignment, back waist gaping, and side seam imbalances add up as the major issues while making bodices. However, the methodical approach to their adjustment and correction leads to successful results. Each problem brings in the need for expertise that covers both the situation at hand and the scale of solutions, merging technical actuality with artful techniques adjustments.

Incorporating Advanced Design Elements:

Through the intricate realm of apparel construction, the transformation from a mundane bodice to a garment with a sophisticated advance elevation is both a practical and a creative task. This exploration considers the delicate art of adding pleats, tucks, and princess seams and adjusting fittings to produce a decorative component of the bodice which will not only create aesthetic appeal but will also demonstrate a certain level of fit [9].

The Art of Pleating: Adding Dimension and Movement

Pleats are a versatile design element that can add volume, texture, and visual interest to a bodice. They range from sharp knife pleats that add sleek lines to a garment, to soft box pleats that introduce a voluminous, flowing effect. The key to successfully integrating pleats lies in understanding their impact on the garment's overall silhouette and ensuring they complement the wearer's body shape.

Incorporating Pleats into the Bodice: To integrate pleats, start by deciding on their placement, width, and direction. For instance, small knife pleats around the waist can create a cinched effect, enhancing the silhouette. The process involves marking the pleat lines on the bodice block, ensuring they are evenly spaced and aligned. The fabric is then folded along these lines and secured, either by stitching the top of the pleats or tacking them at regular intervals.

Critical Analysis: While pleats add a dynamic element to the bodice, they also introduce additional volume, which can be problematic for certain body types. To mitigate this, consider the pleat size and placement carefully, opting for smaller pleats or positioning them in areas that benefit from added fullness. Additionally, the fabric choice plays a crucial role; stiffer fabrics can hold sharp pleats well but may add unwanted bulk, whereas softer fabrics offer a more subtle, draped effect.

Mastery of Tucks: Subtle Shaping and Detail

Tucks, similar to pleats but smaller and sewn down, offer a more subtle method of adding texture and shaping to a bodice. They can be strategically placed to contour the bodice closely to the body's shape, creating a fitted silhouette without the need for darts.

Implementing Tucks: To add tucks to a bodice, first determine their desired effect—whether to introduce decorative texture or to achieve a shaping effect. Tucks can be evenly distributed across the bodice or concentrated in specific areas, such as the waist or bust, to enhance the garment's fit. Each tuck is marked on the bodice block, then sewn down along its length, effectively drawing in the fabric for a closer fit.

Critical Analysis: The primary challenge with tucks lies in their dual nature; they must be precise and uniform to achieve the desired aesthetic and functional effect. Inaccurate sewing can lead to a bodice that looks unintentionally asymmetrical or misshapen. To counteract this, meticulous marking and careful sewing are paramount. Additionally, the choice of fabric is critical, as too thick a material can result in a stiff, restrictive bodice, while too thin a fabric may not hold the tuck's shape effectively.

The Elegance of Princess Seams

Princess seams offer a method of shaping the bodice that is both elegant and highly effective. Running from the shoulder or armhole down to the waist or hem, these seams contour closely to the body's curves, eliminating the need for horizontal darts.

Creating Princess Seams: The process involves adjusting the bodice block by drawing lines where the princess seams will be placed, usually curving slightly over the bust and tapering into the waist. The block is then cut along these lines, and the pieces are shaped and sewn together, allowing for a smooth, fitted silhouette that can easily be adjusted for individual body shapes.

Critical Analysis: While princess seams are highly versatile and flattering, their success hinges on precise execution. Incorrectly shaped seams can distort the bodice's fit, emphasizing areas in an unflattering manner or creating fit issues. To ensure a seamless fit, frequent fittings and adjustments may be necessary, particularly around the bust and waist areas, to refine the seam's curvature and ensure it aligns perfectly with the wearer's body contours.

Conclusion: Balancing Aesthetics with Fit

Integrating advanced design elements like pleats, tucks, and princess seams into a bodice elevates the garment from basic to bespoke. Each technique offers unique opportunities to enhance the garment's visual appeal while addressing fit and silhouette considerations. The key to success lies in a meticulous approach to planning, drafting, and executing these elements, always with an eye toward how they will interact with the body's shape and the chosen fabric's behavior. By critically analyzing the potential challenges these design elements may introduce and employing strategic solutions, designers can ensure that the bodice not only looks exquisite but also fits impeccably, embodying the perfect union of form and function. This journey from a simple bodice block to a complex, elegantly designed piece showcases the depth of skill, creativity, and technical knowledge required in the art of garment construction.

Application and Case Studies

To illustrate the versatility and utility of the bodice block in fashion design, let's revisit the applications and case studies with fictional designer names to bring the narrative to life, while respecting the creative and intellectual property of actual designers.

The Essence of the Bodice Block in Design

The bodice block, a quintessential tool in the arsenal of fashion designers, provides a solid foundation from which a wide range of garments can be created. Its adaptability is key to transforming basic ideas into tangible, wearable art.

Transforming Tradition: Aria Nguyen's Fusion Wear

Aria Nguyen, a designer celebrated for her innovative fusion wear, adeptly combines traditional fabrics and motifs with contemporary silhouettes using the bodice block. Nguyen's technique involves intricate pleating and dart manipulation to seamlessly blend elements from her Vietnamese heritage into modern women's apparel. This not only showcases the bodice block's capacity for versatility but also emphasizes its role in preserving cultural identities within the framework of modern fashion [10].

Innovation Through Adaptation: Julian Marks' Adaptive Clothing Line

Julian Marks has carved a niche in the fashion industry by specializing in adaptive clothing for individuals with physical disabilities. Marks modifies the traditional bodice block to incorporate functional features such as magnetic closures and adjustable seams, making fashion accessible and inclusive. This innovative use of the bodice block highlights its importance in creating garments that cater to a diverse range of body types and needs, further proving its indispensability in design.

Sustainable Practices in Design: Eliza Chen's Eco-friendly Collection

In the realm of sustainable fashion, Eliza Chen stands out for her commitment to environmentally friendly design practices. By optimizing the use of the bodice block, Chen minimizes fabric waste through meticulous pattern placement and zero-waste design techniques. Her approach not only results in stylish, well-fitting garments but also aligns with the pressing need for sustainability in the fashion industry, demonstrating the bodice block's role in promoting eco-conscious design.

Bridging Technology and Tradition: Max Harper's Digital Innovation

Max Harper represents the cutting edge of fashion design, merging traditional techniques with digital technology. Harper utilizes digital pattern-making software to transform the conventional bodice block into 3D models, facilitating virtual fittings and adjustments. This blend of technology and tradition streamlines the design process, reduces waste, and exemplifies the modern applications of the bodice block, ensuring its continued relevance in the digital age [11].

Conclusion

Through the inventive works of Aria Nguyen, Julian Marks, Eliza Chen, and Max Harper, the enduring relevance of the bodice block in fashion design is unmistakably highlighted. From traditional garment construction to the forefront of digital design innovation, the bodice block remains a fundamental element, enabling designers to explore new territories, meet diverse needs, and address contemporary issues. Its adaptability, precision, and utility make it an irreplaceable component of the design process, bridging the past, present, and future of fashion.

Future Trends in Bodice Development

The exploration of the bodice block, particularly in the context of its future trends and innovations, invites a deep dive into the evolving landscape of fashion design. This foundational element of pattern making, which has long served as the cornerstone for creating well-fitted and expressive garments, stands on the brink of transformative changes influenced by new technologies, sustainable practices, and emerging trends. Let's delve into these potential future directions and how they may redefine the use and application of the bodice block in pattern making.

Technological Integration in Bodice Design

The advent of technology in fashion design, particularly through the incorporation of 3D printing and digital modeling, offers exciting prospects for the evolution of the bodice block. Imagine a scenario where designers, equipped with sophisticated software, can digitally sculpt the bodice block around a virtual mannequin that precisely mimics the client's measurements. This not only streamlines the design process but also allows for an unprecedented level of customization and precision in fit.

Moreover, 3D printing technologies hold the potential to revolutionize fabric creation and garment construction. Designers could potentially print a bodice complete with integrated structure or embellishments directly onto fabric, challenging traditional sewing and construction techniques and opening up new avenues for creativity and innovation in bodice design.

Sustainable Practices Shaping the Future

Sustainability is a pressing concern in the fashion industry, urging designers to reconsider every aspect of garment creation, including the use of bodice blocks. Future trends may see the bodice block being utilized to minimize fabric waste through zero-waste pattern making. Designers like Eliza Chen, who prioritize eco-friendly practices, are leading the way by developing innovative methods to lay out patterns, including bodice blocks, in a manner that ensures every inch of fabric is used effectively, leaving negligible waste. Furthermore, the application of sustainable materials in bodice construction—such as recycled fabrics or biodegradable textiles—complements the eco-conscious use of bodice blocks, highlighting a holistic approach to sustainability in fashion design.

Emerging Trends Influencing Bodice Block Usage

The fashion industry is perennially evolving, with trends shaping and reshaping the contours of garment design. The bodice block, traditionally seen as a static element in pattern making, is now subject to reinterpretation and reinvention. Designers are exploring beyond the conventional boundaries, incorporating elements like modular designs or interchangeable parts into the bodice, reflecting a growing demand for versatility and adaptability in clothing.

Another emerging trend is the fusion of technology and wearables, where bodice designs could integrate smart fabrics or electronic components, allowing garments to interact with the environment or monitor the wearer's health. This intersection of fashion and technology challenges pattern makers to think of the bodice block not just as a form-fitting element but as a dynamic platform for innovation.

The Enduring Significance of the Bodice Block

Despite these forward-looking trends and innovations, the essence of the bodice block as a tool for achieving the perfect fit and expressing creativity remains unchanged. Its adaptability and versatility ensure its continuing relevance in the fashion industry, serving as a testament to the skill, precision, and artistic vision of designers and tailors.

As we speculate on the future of bodice block design and usage, it becomes clear that its potential is boundless. The integration of new technologies, a commitment to sustainability, and the embrace of emerging trends are not just redefining the bodice block's application but also reinforcing its fundamental role in the art and science of pattern making.

Conclusion

As we draw this exploration to a close, the journey through the intricacies of the bodice block underscores its pivotal role in the tapestry of garment construction. This foundational element, a keystone in the arch of fashion design, embodies the marriage of precision and creativity that defines the craft. From the initial sketches to the final stitches, the bodice block serves as both a map and compass, guiding designers through the complex landscape of fabric and form.

This chapter has navigated through the practical applications and transformative possibilities of the bodice block, revealing its versatility and adaptability. Through the lens of technology and innovation, we have glimpsed a future where the bodice block transcends its traditional boundaries, merging with digital tools and sustainable practices to forge new paths in garment design. Yet, amidst these advancements, the essence of the bodice block remains unchanged—a testament to the enduring artistry and skill of pattern making.

The case studies and examples provided herein are not just narratives; they are beacons of inspiration, showcasing how the bodice block can be manipulated, reimagined, and refined to create garments that resonate with both the wearer and the times. Designers like Aria Nguyen, Julian Marks, Eliza Chen, and Max Harper exemplify the boundless creativity and innovation that the bodice block facilitates, each bringing their unique vision to life while addressing contemporary challenges in fashion.

As "Precision and Grace: The Art of English Pattern Making in Garments" reaches its conclusion, it is clear that mastering the bodice block is both a rite of passage and a continuous journey for designers. It demands a deep understanding of human anatomy, a keen eye for detail, and an open heart for exploration. Whether adapting the bodice for technological integration, championing sustainability, or embracing emerging trends, the bodice block stands as a canvas ready to be painted with the broad strokes of innovation and the fine lines of tradition.

In the realm of fashion design, where the only constant is change, the bodice block remains a steadfast ally, empowering designers to navigate the evolving demands of fit, function, and aesthetics. It is a bridge connecting the past with the future, tradition with innovation, and art with science. As we look forward to the next chapter in garment construction, the bodice block will undoubtedly continue to play a central role, inspiring new generations of designers to push the boundaries of creativity and craftsmanship.

In essence, the bodice block is not merely a tool for pattern making; it is a symbol of the fashion industry's perpetual quest for perfection, a quest that balances the grace of form with the precision of fit. It encapsulates the spirit of garment construction, inviting all who dare to dream, design, and redefine the fabric of our world.

References

1. Y. Zhao, F. Dai, M.M. Gupta, W. Zhang, Ontology modeling for intelligent computer-aided design of apparel products. Int. J. Autom. Technol. **10**(2), 144–152 (2016). https://doi.org/10.20965/ijat.2016.p0144
2. P.E. Zwane, N.A. Magagula, Pattern design for women with disproportionate figures: a case study for Swaziland. Int. J. Consum. Stud. **31**(3), 283–287 (2007). https://doi.org/10.1111/j.1470-6431.2006.00549.x
3. A. Piper, K. Townsend, Crafting the composite garment: the role of hand weaving in digital creation. J. Text. Des. Res. Pract. **3**(1–2), 3–26 (2015). https://doi.org/10.1080/20511787.2015.1127037
4. K.A.R. Richards, M.A. Hemphill, A practical guide to collaborative qualitative data analysis. J. Teach. Phys. Educ. **37**(2), 225–231 (2018). https://doi.org/10.1123/jtpe.2017-0084
5. M. Ahmed, T. Alrushaydan, S. Gill, S.G. Hayes, K. Brubacher, The suitability of body scanning measurement in pattern drafting methods, in *Proceedings of 3DBODY.TECH 2019—10th International Conference and Exhibition on 3D Body Scanning and Processing Technologies, Lugano, Switzerland, 22–23 Oct. 2019*, Ascona, Switzerland: Hometrica Consulting—Dr. Nicola D'Apuzzo, Oct. 2019, pp. 58–67. https://doi.org/10.15221/19.058
6. L. Griffin, C. Compton, L.E. Dunne, An analysis of the variability of anatomical body references within ready-to-wear garment sizes, in *Proceedings of the 2016 ACM International Symposium on Wearable Computers* (ACM, New York, NY, USA, 2016), pp. 84–91. https://doi.org/10.1145/2971763.2971800
7. A. Foerster, R. Gray, R. Cañal-Bruland, Size estimates remain stable in the face of differences in performance outcome variability in an aiming task. Conscious. Cogn. **33**, 47–52 (2015). https://doi.org/10.1016/j.concog.2014.12.004
8. Y.J. Kim, D. Koo, Enhancing bra design for post-mastectomy patients: incorporating MRI data and innovative textiles for optimal support and comfort. J. Eng. Fiber. Fabr. **19** (2024). https://doi.org/10.1177/15589250241241752
9. M.S. Hassan, S. Salawdeh, J. Goggins, Advanced finite element simulation of ductile structural steel incorporating a crack growth model. Structures **15**, 94–114 (2018). https://doi.org/10.1016/j.istruc.2018.06.002
10. T.-C. Truong, Q.B. Diep, Technological spotlights of digital transformation in tertiary education. IEEE Access **11**, 40954–40966 (2023). https://doi.org/10.1109/ACCESS.2023.3270340
11. P.V.M.V.D. Udovita, Conceptual review on dimensions of digital transformation in modern era. Int. J. Scien. Res. Publ. (IJSRP) **10**(2), p9873 (2020). https://doi.org/10.29322/IJSRP.10.02.2020.p9873

Chapter 7
Darts, Pleats, and Gathers

Rida Alvi, Zafar Javed, Nazish Huma Khan, and Gohar Ali Hayat

Abstract Although the human body is comprised of so many different forms and structures, pattern makers, and designers find it challenging to create accurate patterns. It's crucial to understand fundamental body measurements and precise processes in order to create a dress-fitting, stylish, and comfortable appearance. Clothing used to be used merely to cover the body in earlier times. Nevertheless, as time goes on and the fashion industry develops, people are becoming more aware of their body types and seeking out their own distinct looks. To this end, darts, dart manipulation, pleats, and gathers are used to give clothing an attractive, fit, and full appearance. At first, fittings were done with darts, but now that Techniques for manipulating darts have been created, and unique, imaginative cutlines are displayed. Based on conventional measurements, this section includes the intake, placement, and location of the dart. In addition, dart manipulation provides a style line and gives the striped pattern a new depth; the garment's fit is unaffected by these manipulations. Throughout history, pleats have been regularly employed to create surfaces, enhance fullness, and provide a quick way to add fittings. Pleat is an essential phrase in sewing and fashion, as it refers to a type of fold that gives fabric structure, shape, and adornment. Pleats can be affixed permanently to a garment to add volume and texture. Cloth is folded, then sewn or pressed together to form pleats. Appeal fulfilled. To create fullness, gather a piece of fabric and bundle it to fit onto a smaller, premeasured length of cloth. Usually, gathering is done by using a longer machine stitch that enables the fabric to be gathered by hand by pulling the threads. A third or half

R. Alvi
The Government Sadiq College Women University, Bahawalpur, Pakistan

Z. Javed
School of Art and Design, National Textile University Faisalabad, Faisalabad, Pakistan

N. H. Khan
Department of Fashion Design, NFC University Multan, Multan, Pakistan

G. A. Hayat (✉)
Department of Industrial Design, School of Art and Design, Wuhan University of Technology, Wuhan, China
e-mail: gohar.ali@iub.edu.pk

M. Q. Khan et al. (eds.), *Garment Sizing and Pattern Making*, SDGs and Textiles,
https://doi.org/10.1007/978-981-97-7683-2_7

of the original width is typically used for gathering, depending on the fabric and the desired level of fullness.

Keywords Dart · Dart manipulation · Pleats · Gathers fashion design

Introduction

When discussing the making of clothing, it seems so apparent to work on different approaches and techniques linked to pattern cutting and body-type-appropriate shape; darts, pleats, and gathers immediately spring to mind for this reason. For fitting purposes and creative design of the garment's cutlines, construction, and fullness, these structural components are crucial. These components become crucial for dress-makers and fashion designers to understand during cutting. A dart is a fold or tuck point in fabric that gives an article of clothing ease and shape. A triangle-shaped inlet called a dart is sewed into a garment to give it a three-dimensional shape, particularly to accentuate a woman's bust. Darts are a necessary tool for manufacturing dresses. It can give two-dimensional, flat fabric an organic shape. Only the body type and necessary level of fitness will define the dart's position and quantity. The shape that is inserted at the neck, shoulder, bust, and waistlines by tapping at both ends determines the dart's width and length. With their manipulation skills, darts contribute significantly to the creation of creative patterns and cutlines for dressing, regardless of the fitting objective [1].

Pleats, on the other hand, are folds in the cloth that can resemble an accordion. Pleats are available in various sizes and shapes. While stitched pleats are regarded as tucks (used for ornamental purposes) and are another swing technique gathering that is employed as bundling up smaller pieces of fabric to create fullness, pressed, and ironed pleats give the cloth structure. The waistline, shoulder seams, ruffles, yokes, and sleeve cuffs are the main areas where gathering is observed [2].

History of Dart and Pleat

Following World War II, people started to pay more attention to clothing and fashion. The French fashion designer Pierre Alexandre was the first to employ dart in the early 1800s. He made women's clothing with darts at the breast and waist. Darts were originally employed in menswear mostly for fitting and to contour items that appeared flat. Darts had no significant role in the early days because pattern cutting was difficult to do. The use of darts in women's clothing pockets was invented in the early 1950s by the New York City firm Evan-Picone. In order to reduce the possibility of rips or tears, the darts help keep the pocket open and therefore more easily accessible. Pleats were originally used to indicate class distinction in ancient Egyptian tunics. They preferred to impose pleats using linen fabric. The secret of how the Egyptians achieved pleat effects remains a mystery. The archaeology proposes two methods: the cloth is folded in the appropriate pattern, knotted, and wetted. The pleats would eventually show up after the fabric had dried and the ties were removed. Using glue and sizing material to secure the cloth to a wooden board in order to create pleats is a significant alternative. Making pleats for greater rankings in history is the result of this extraordinary process. Pleats were utilized to decorate collars in the sixteenth century, and these garments were kept in specific boxes to maintain their shape.

Famous fashion designer Mariano Fortuny developed novel pleating techniques in the 1900s, utilizing them to produce the renowned Delphos garment. The silk fabric that made up the Delphos dress's signature fit the body's curves with ease. On this cloth, 450 folds in all were accomplished. Pleats vanished rapidly unless the garment was folded into a ball and laundered, in which case they remained intact. The most well-known theater actress of the 1920s wore this dress. Another sewing method that has been used since ancient times to offer clothing flare, volume, and ease is gathers. Gathers were generally used around the waist and sleeves of clothing to create surface details and aesthetic appeal [3].

What is Dart and Dart Manipulation

Dart

Darts are defined as folds sewed into fabric that give an article of clothing a three-dimensional shape. To give a garment ease and shape, darts are folds that are sewed into the cloth. They are commonly utilized in women's apparel to fit the wearer's figure. The best approach to shape a garment and get rid of extra fabric to create a unique shape is with darts. After cutting away any extra material, the right side of this triangle-shaped cut is sewed together. You can see a narrow cone as you draw the wedge together. This cone point is quite clear on paper, but because the fabric is more flexible, it is moved about the peak of the cone. In a garment, the dart ends at the full area of the body. The tip and the intake are the two main components of a dart in a basic flat design. The intake denotes the quantity of cloth removed, and the tip is where the dart is aimed. This input is exactly proportionate to the angle that the dart deducted from the pattern [3].

Components of Darts

Abbreviations and Measurements for Basic Bodice

The two dart legs should appropriately overlap on top of one another during folding. The dart legs provide a stitch line to follow throughout the dart-building stage since they are transferred from the pattern onto the fabric. The center line, also called the

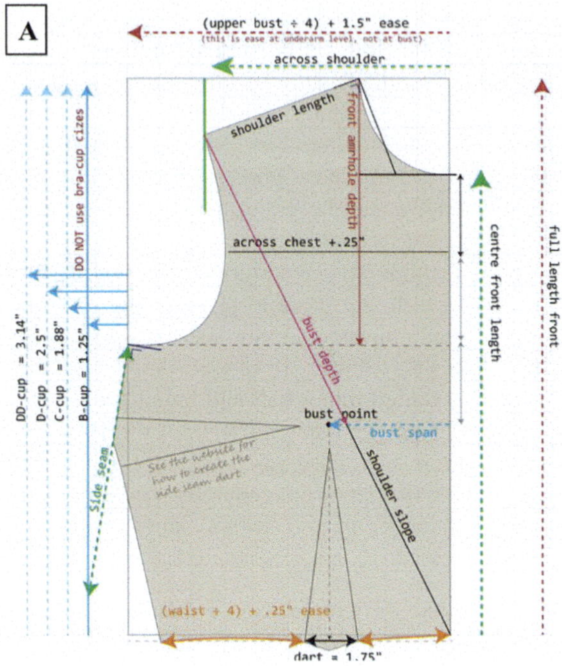

Fig. 7.1 Basic front bodies pattern

dart fold line, is positioned equally between the two dart legs and runs down the middle of the dart. The two dart legs are directly aligned when the dart is folded along this fold line, which facilitates easier folding. The angular point where the center line and the two dart legs cross is known as the vanishing point. It is also the point where the dart disappears into the smooth surface of the fabric.

Here, the basic bodice blocks front and back, are labeled in the figures (Figs. 7.1 and 7.2). All measurements are shown in inches [4].

Anatomy of Dart

There are four parts of a dart:

- Dart point.
- Dart legs.
- Dart intake.
- Dart root (Fig. 7.3).

Three major types of darts are widely used in the dressmaking process.

Single-Pointed Darts

In dressmaking, single-point darts are most frequently employed. They have a single vanishing point and always extend from a seam. The seam edge that continues into the seam allowance is the dart's widest point. Single-pointed darts have two dart legs and a center fold line much like any other dart. The broadest portion is always sewed

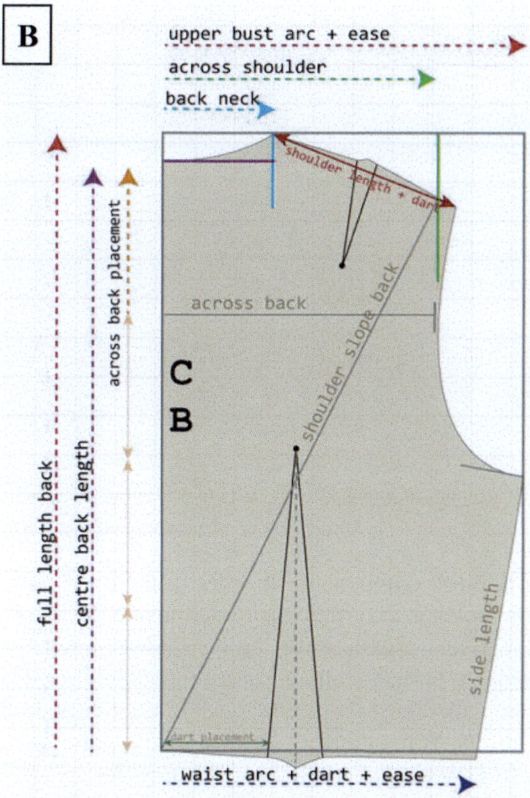

Fig. 7.2 Basic back bodies pattern

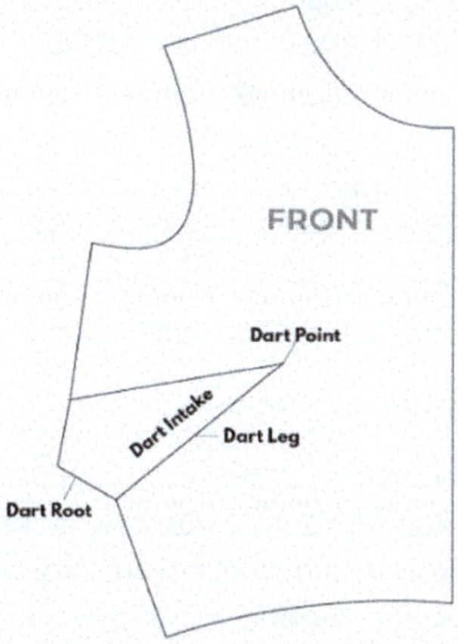

Fig. 7.3 Anatomy of dart

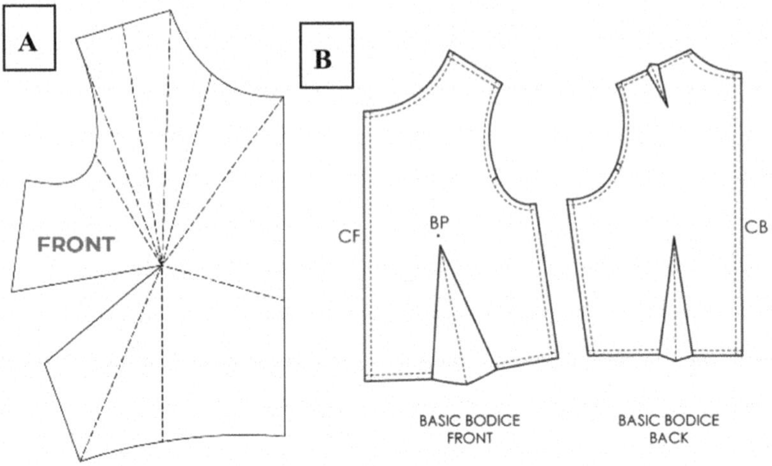

Fig. 7.4 Single-pointed dart

to the thinnest component. In order to facilitate the dart transfer and construction process, notches are often positioned inside the seam allowance at the end of each dart line. Because of their versatility, single-pointed darts can be positioned vertically, horizontally, or diagonally on an article of clothing. The most popular single-pointed darts are called bust darts, and they are used only on the front of clothing, arranged horizontally. Waist darts, or vertical single-point darts, are typically located on the sides, front, and back of clothing [5] (Fig. 7.4).

French Darts

French darts are exclusively used on the front of a garment and are less popular in dressmaking. Usually located two to three inches above the waist, this dart pattern extends diagonally from the seam to the hip or bust.

French darts might be curved slightly or they can be straight. Because the excess needs to be cut and sometimes snipped for tension release (on curved French darts), they require a different construction technique. It's frequently used on silk tops for a more polished appearance (Fig. 7.5).

Contour Darts

Known by other names like fisheye darts or double-pointed darts, contour darts are used on tunics, blazers, jackets, and fitted dresses. They offer two vanishing points that stretch in opposite directions and a tailored intake at the waist. When two waist single-pointed darts are linked at their widest ends, the result is a contour dart. The top and bottom vanishing points of front contour darts taper to the breast and hip, respectively.

Alternatively, back contour darts feature a bottom tip that taper toward the buttocks and a top dart point that is positioned at the broadest part of the back. Uncertainty will result from a too wide dart intake. To reduce tension in this instance, clip it

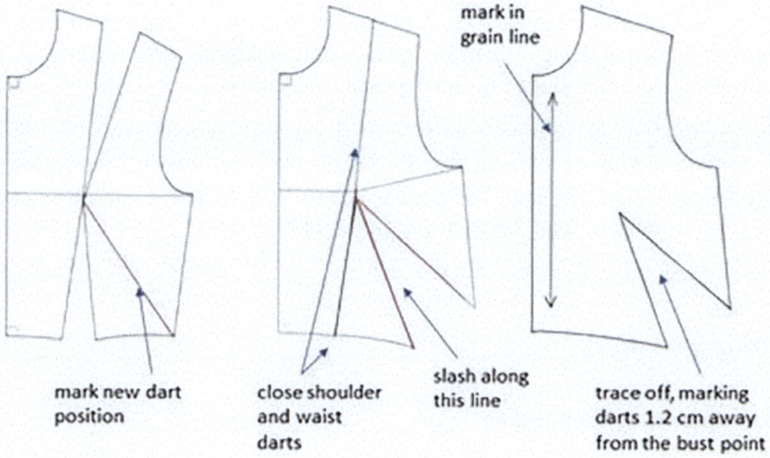

Fig. 7.5 French dart

horizontally at the waist, which is its widest position. This procedure is not usually required, particularly for items that are unlined and have obvious dart surplus within. Use the clipping process if the finished contour dart does not lay flat [6] (Fig. 7.6).

Dart Names and Style

Two types of darts are frequently seen in women's blouses:

Vertical darts: Sewn from the blouse's bottom to a point centered around the bustline. The seamstress uses this dart, which is typically visible at the front of an item of clothing, to draw the blouse's bottom in toward the wearer's waist.

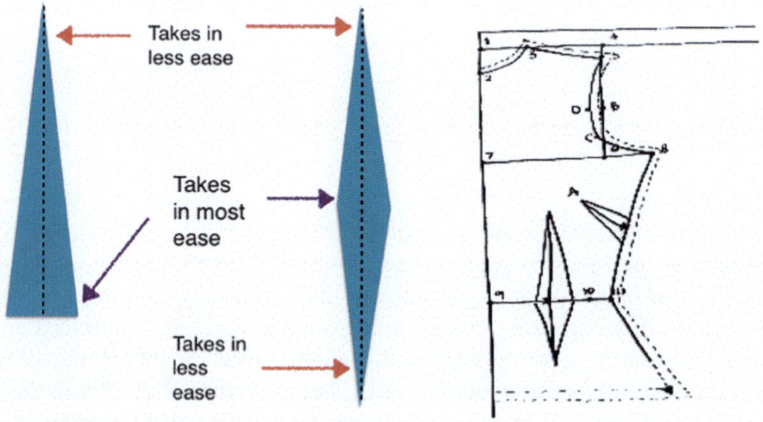

Fig. 7.6 Contour dart

Bust darts: The fabric beneath the breasts can fit closer to the wearer rather than hanging because of these little triangle folds that make space for the breasts. Darts for the waist and bust lines fall into several categories (Table 7.1).

These are the typical front body darts; each dart is measured starting from where it is on the bodice and ends 1″/2.5 cm from the bust peak. Additionally, seam allowances are as follows: 1/2 or 1/4 inch for regular seams, 1/4 inch for curved seams like armholes and necklines, and 3/4 inch for zippers [7].

Dart Manipulation

Dart Manipulation Techniques

The skill of moving darts around a pattern to create art is known as dart manipulation. Usually, women's clothing has it on top. The skill of shifting the dart by folding and shutting the undesirable dart to create a new opening for fitting and creatively creating new patterns. It offers designers countless options and is the most versatile aspect of pattern cutting. You can make pleats, gathers, or style lines out of darts. Their placement on the body is crucial; these methods alter the style and appearance of the garment in addition to creating fit, shape, and volume.

Techniques to achieve dart manipulation are as follows:

Pivot Technique

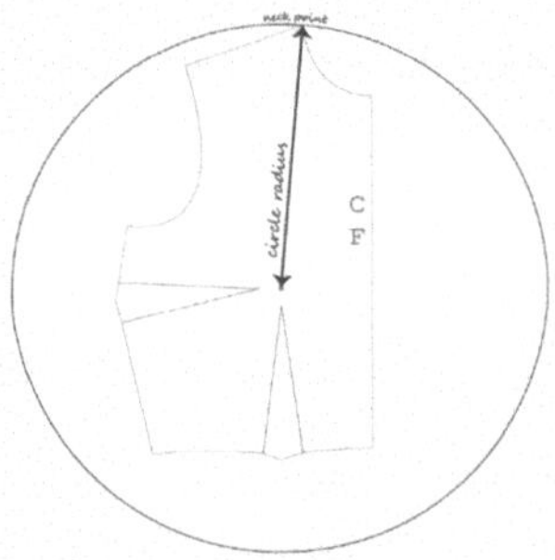

The pin and pivot dart rotation method requires tracing a new design from the original. To trace the new design, first set the pattern with the dart to be rotated on top of a different sheet of paper. A pin is put into the dart point to hold it in place. Next, a random section of the original design and the original dart's leg are traced onto the paper. When the dart reaches the edge of the pattern piece, the tracing moves on to a new spot, starting with one dart leg. The other dart leg is then aligned with the traced dart leg by rotating the pattern around the pinned dart point. Then, starting from the same location on the initial pattern, tracing can be continued. After that, the pattern is

Table 7.1 Types of darts

Basic darts	Description	Images
French dart	The dart starts at the side seam and ends below the bust point	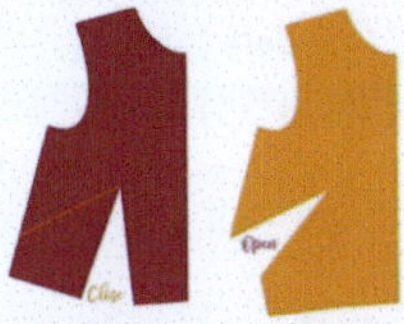
Armhole dart	It begins from the armhole pitch point toward the apex. It is used to shape tops and blouses around the bust point	
Mid shoulder dart	This dart is positioned from the princess seam at the shoulder to the bust point	
Center front neck dart	It's close at the waist opens at CF's neck and is a diagonal dart in front of the center. It is used to suppress the fabric	
Center front bust dart	Opening at center front and close at waist. It creates little gathers at CF. It is also used to form a bib or yoke effect	
Center front waist dart	Close at the waist and open at the center front waist. It is needed to contour the waistline	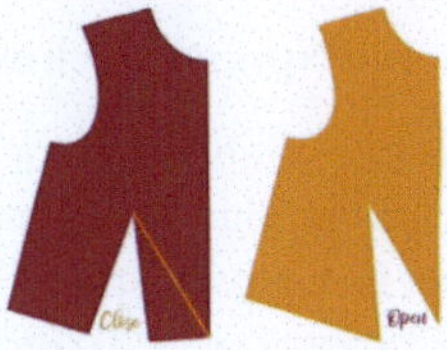

(continued)

Table 7.1 (continued)

Basic darts	Description	Images
Shoulder tip dart	It transfers from the waist to the shoulder tip. This dart is tricky	
Mid neck dart	From the mid-neckdown to the bust point. It adds fullness to a pattern and opens up the neckline	
Straight side seam dart	It's closed at the waist and open at the side seam. Horizontal dart from the side seam	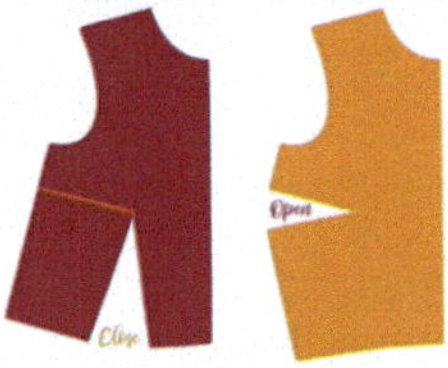

eliminated, and the newly drawn dart legs are placed between the pinhole-designated dart point and the space created by the pattern's rotation.

Usage of Pivot Technique

Process of Pivot Dart Technique

- Shifting a dart from one spot on a pattern to another.
- Breaking apart a large dart into many smaller ones.
- Gathers, pleats, and tucks can be produced out of the fabric gathered in a dart.

Begin by tracing one part of the bodice. Tracing clockwise from the bust point to the end (Fig. 7.7):

Slash and Spread Technique

The design is chopped, or slashed, to the dart point. To enhance fullness, spread the necessary amount, or remove fullness, move closer together. A flat-pattern method approach that uses spreading and cutting to adjust the sloper in order to achieve the desired appearance. To move bust darts from one section of the body to another, close the previous dart to the bust point and slash to the bust point on the new dart line. The new dart will be delivered when the cut opens.

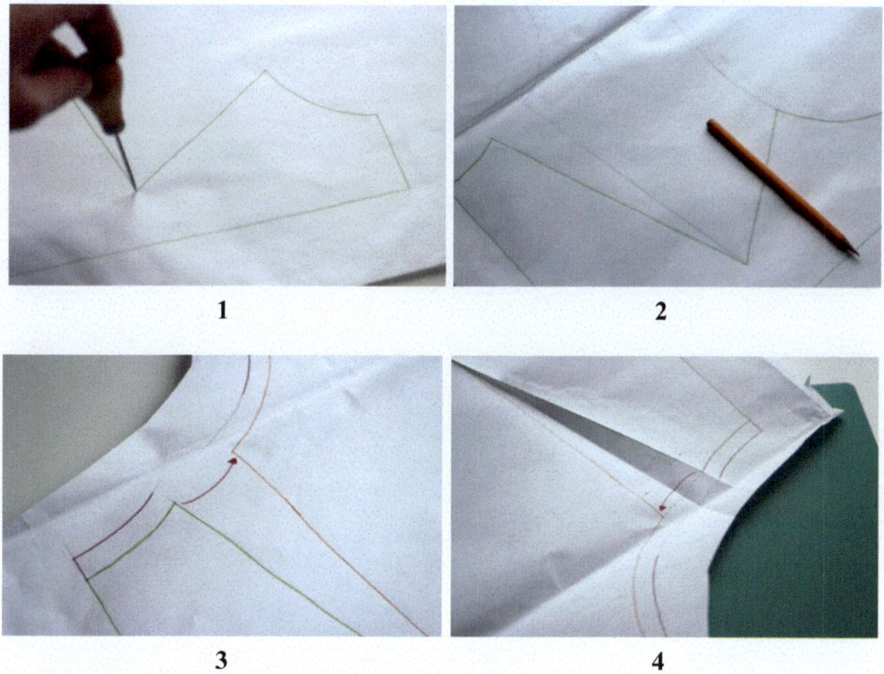

Fig. 7.7 Process of making pivot dart

Process of Slash and Spread Dart Technique

(a) Create a copy and choose the dart location.
(b) Fold the dart and mark it with a tracer.
(c) Making new darts and cutting to attain proper shape (Fig. 7.8).

Ideas Generated by the Dart Manipulation Technique

The creative ideas will be more possible and the clothing will look more appealing with the darts moved on the bust area. These are some illustrations of patterns that demonstrate how darts are inserted and alter the garment's cutlines. In addition to applying symmetrical and asymmetrical manipulation, gathers added to the dart are displayed as blows. The line drawings below depict a partially frontal body segment, with a dart symbolized by the green mark [8] (Figs. 7.9 and 7.10).

There will be additional options for manipulating the dart now that the entire front piece is being worked on rather than just half of it; it can add asymmetrical designs and style lines. The darts can be made to disappear by passing between the dart points and drawing any form. A line that meets the bodies' edges at the same location is necessary to create this shape. The pattern won't lay flat if it doesn't. Cuts, or notches, are required (Figs. 7.11 and 7.12).

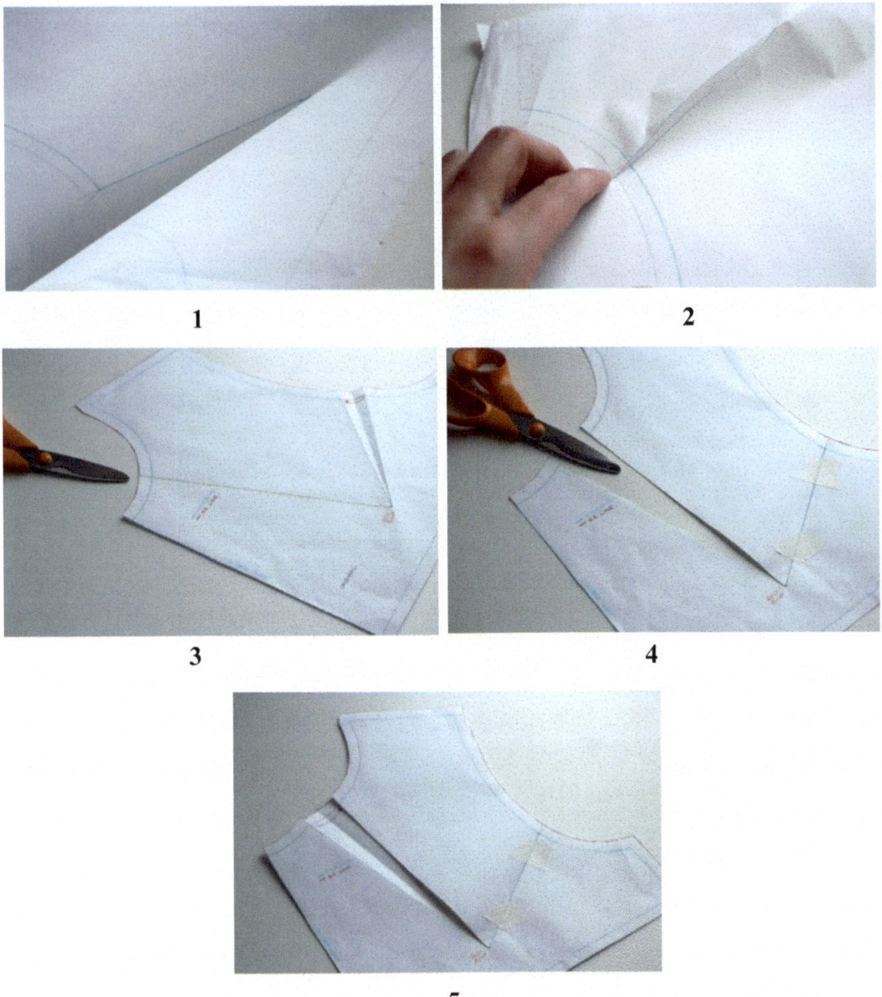

Fig. 7.8 Slash and spread technique step by step

Darts can turn into tucks or pleats. These sorts of darts give volume to the garment and don't give extreme fitness to the body. Some of its examples are shown below (Figs. 7.13 and 7.14).

Dart Manipulation on Skirts

Just like with the bodies, drafting can also be applied to the bottom. You can adjust the direction of the darts to any direction. In the dart point and the skirt's pivot point are the same. A table with a few instances is displayed (Table 7.2):

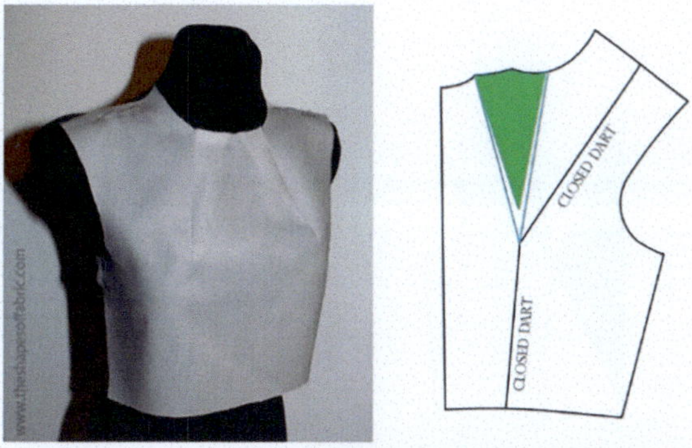

Fig. 7.9 Symmetrical dart at neckline

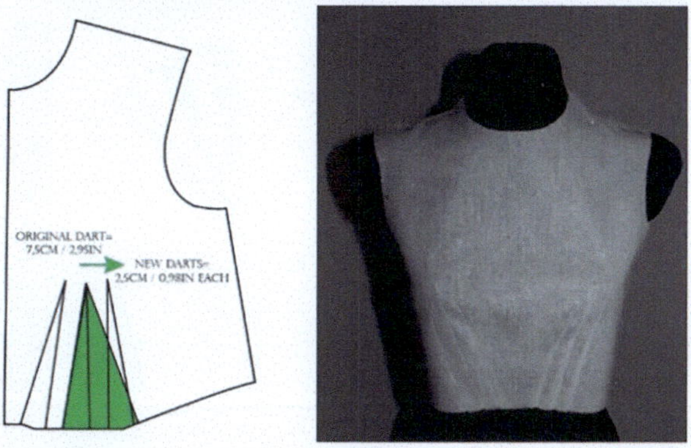

Fig. 7.10 Symmetrical dart at waistline

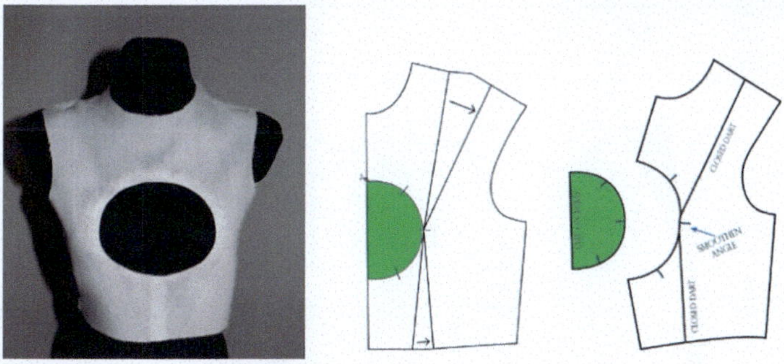

Fig. 7.11 Asymmetrical dart circular shape

Software is Used for Pattern Making

In order to reduce time and fabric waste during mass production, patterns are drawn in the program. Pattern-making software includes Style CAD, TUKAcad, Pattern Smith, Optitex, Modaris Quick Estimate, and others. With just a few keystrokes, these CAD programs can create and manipulate darts, greatly increasing pattern speed and

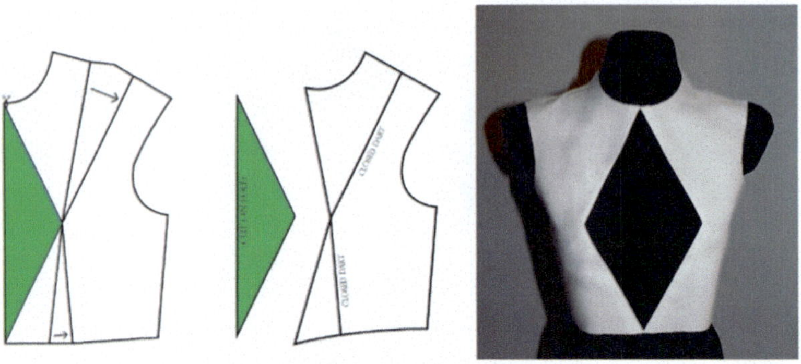

Fig. 7.12 Asymmetrical dart diamond shape

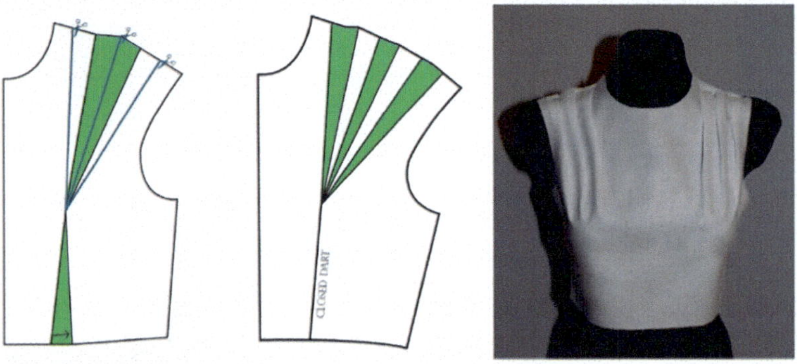

Fig. 7.13 Dart transfer into gathers

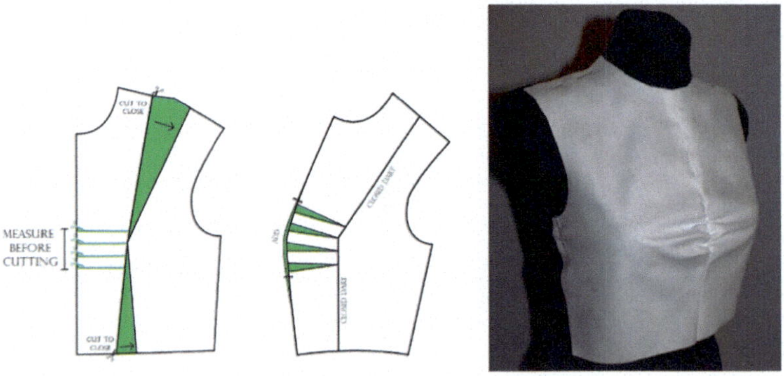

Fig. 7.14 Dart transfer into gathers at center front

Table 7.2 Different types of dart manipulation on skirt

Skirt style	Line drawing	Stitched sample
Basic skirt	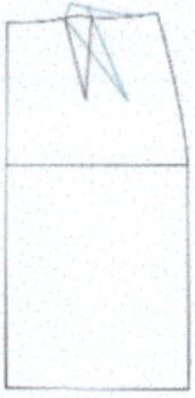	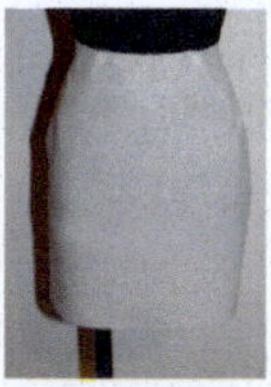
Basic skirt with curve dart	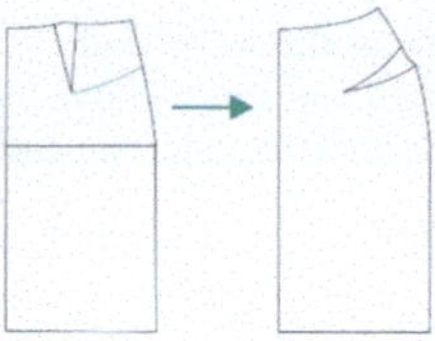	
A-line skirt Remove ¼ of the total circumference of each piece of the colored dart to achieve a fit look	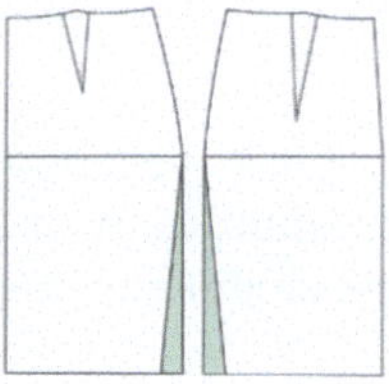	
Close the dart and add 4 cm or 1.5 in on both sides of the hemline. It will give volume	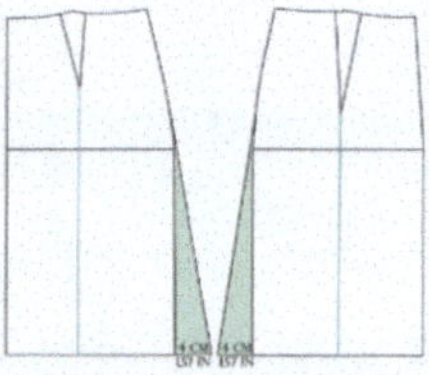	
High waist The rise in the waist is a bit different. The added portion should follow the straight guidelines of the dart	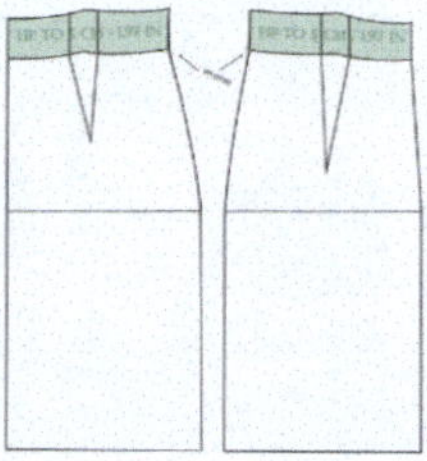	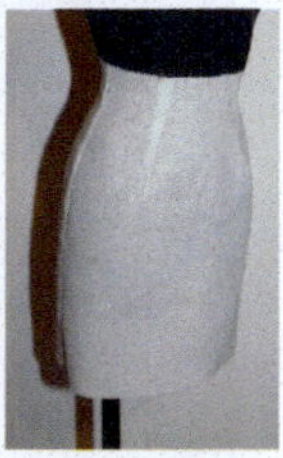

(continued)

Table 7.2 (continued)

Skirt style	Line drawing	Stitched sample
Low waist Lowering the waistline to 5 cm so the dart intake should be 0.5–1 cm	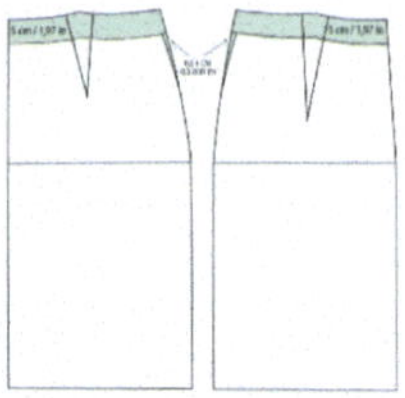	
Adding yoke The yoke line is the same as the waistband. Choose the required height and shape above the hip line and adjust the dart so the yoke can end there	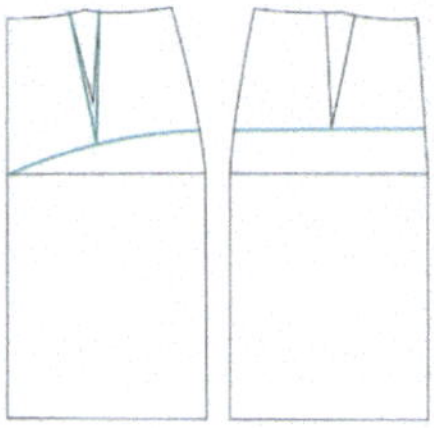	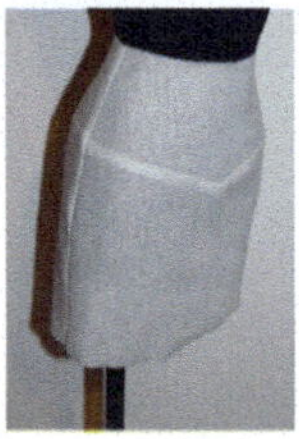

accuracy. With just one click, patternmakers may create patterns in various sizes thanks to the program. Additionally, it can lessen the likelihood of errors; certain software also renders the clothing in three dimensions (Figs. 7.15 and 7.16).

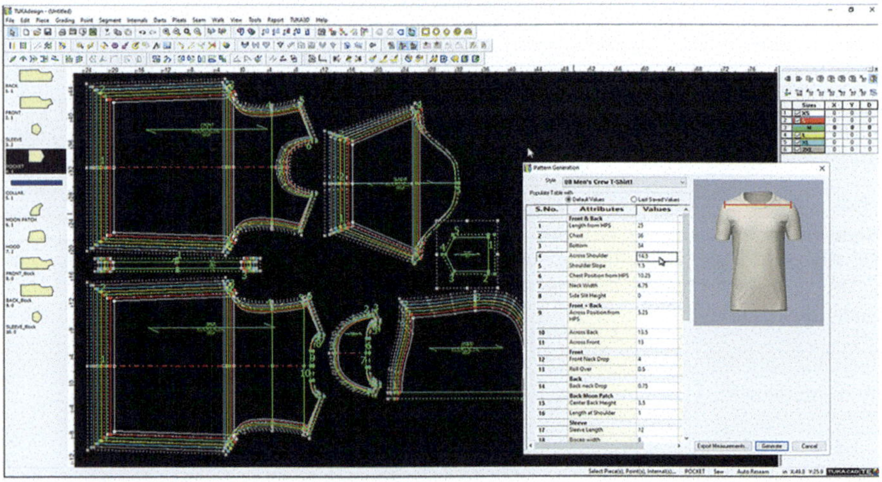

Fig. 7.15 TUKA cad software interface

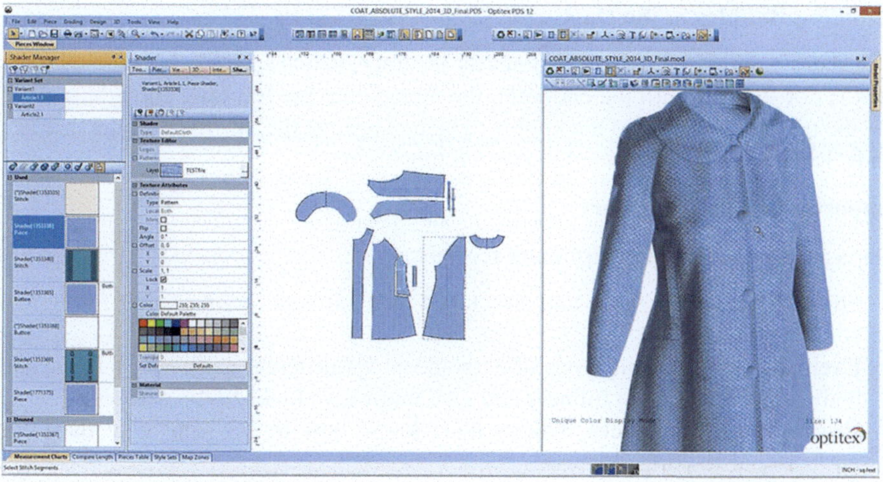

Fig. 7.16 Optitex software interface

How are Darts Sew?

Dart stitching is not the same as regular stitching. To produce a precise, well-fitting pattern while sewing darts, there are a few tricks and methods to know. Every sort of dart is sewed in a particular way. A few of these are explained in detail below [8].

Fabric Preparation

Using a tailor's track or fabric-making technique, transfer all patterns to the wrong side of the material. To create a dart point, bring both legs together to form a center point and the crease.

Fold the Dart in Half

To check the alignment of the dart legs, fold it in half, mark it, and pin it. Then, check the other side of the dart to make sure the pin passes through the opposing dart leg; if not, make the necessary adjustments.

Setting Stitch Length and Length

While the thread length changes according to the fabric weight, the straight stitch is more convenient. For heavier fabrics, a longer stitch length is utilized, and for lighter fabrics, a shorter thread length.

Sew the Dart

Sew toward the point of the dart starting at its widest position. Stitch off the edge of the fabric and tie off the thread by reaching to the dart point. Avoid backstitching since it may result in a lump at the site.

Avoiding Bubbles in the Dart

The sharp point of the dart determines its accuracy, which can only be attained if the dart is not stitched until the endpoint, the thread ends are left long enough to allow for the tying of a knot, and the knot is tightened against the tip. Take off the thread's ends, leaving a 6 to 1.3 cm tail to prevent unwinding.

Importance of Sewing Dart

Darts are essential for three primary reasons:

- **Shape**: Dart added shape and structure to the garment, giving better fitting and comfort to wear.
- **Exclusiveness**: Various types of darts meant for different garments, fabrics, and body mass. Darts form fullness and give innovative looks to the garments.
- **Professionalism**: Pattern makers must know how to sew and take darts for professional-looking clothes.

Dart Equivalents

Gathers, pleats, and tucks are examples of dart equivalents. Style lines are an additional instance.

Pleats

Pleats are a form of fold that are used to provide cloth structure, design, and shape. Pleats is an important term in the sewing and fashion industries. Pleats can be permanently fastened to a garment to add volume and texture. The cloth is folded to create pleats, which are then sewn or pressed together. Tucks is the expression for the stitched pleats. Depending on the intended effect, several sizes of folds can be achieved. Pleats have been used to enhance the look and functionality of clothes for generations. It is typically seen in royal skits, uniforms, and kilts. Pleats give the cloth a well-organized drape, produce movement, and facilitate ease of fit in the garment. however, changes the fabric's appearance and texture. The type of fabric (natural or synthetic) and folding method affect the pleat's texture.

Pleat Role in Fashion Design

Because pleats retain more fabric and require a time-consuming pressing process, the elite class in ancient Egypt fashioned them to display their royalty. It provides the volume, texture, and dimension to the clothing. The development of 3-D volume in the cloth is largely dependent on the dynamic folding technique. Pleats have a utilitarian purpose for airflow and mobility of movement in addition to their aesthetic appeal. Box and side pleats are the most widely used pleat styles. The circular, curtain, draped, fluted, accordion, cartridge, and French are a few more popular varieties. They can be moved, inserted, and turned to both sides (Fig. 7.17).

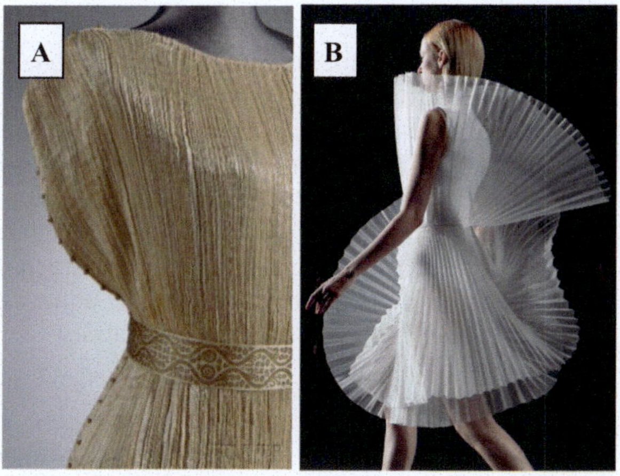

Fig. 7.17 Pleats in clothing

Terminology of Pleats

The pleats folded along a straight line are known as fold line, and the fold placed with another line is called placement line.

- **Pleat Depth**: Distance from the outer edge of the fold to the inner edge of the fold.
- **Pleat Underlay**: The quantity of the fabric folded underneath the pleat. The underlay is always twice the pleat depth (Fig. 7.18).

Function of Pleats

- The size and positioning of the pleats vary to create a distinctive style, depending on the garment designer's vision or context. Here are some typical explanations for why a designer could use pleats in a piece of clothing.

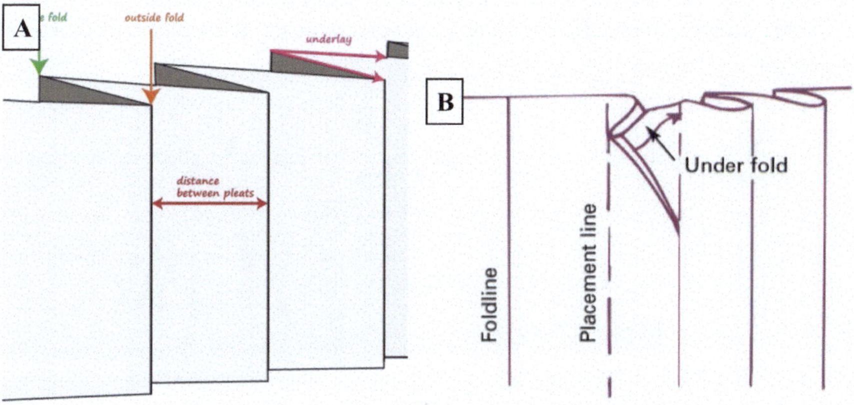

Fig. 7.18 Terminology of pleats

- Fit and Function: Pleats can make clothing more roomy and flexible, which makes them useful. Pleats, for instance, can provide room around the hips and thighs and improve movement in skirts or pants.
- Style and Aesthetics: Pleats are commonly utilized in apparel to provide depth, texture, and visual attractiveness. They are able to evoke a sense of movement, structure, and volume. Pleats can be strategically placed to create a particular silhouette or to highlight particular areas of the fabric.
- Shape and Tailoring: Pleats can give fabric structure and shape, making it appear more fitting and put together. They can be utilized, especially around waistbands and cuffs, to control the fullness of the fabric.

Types of Pleats

Ten different kinds of pleats are typically observed and displayed with figures.

Box Pleats: Because they sculpt the waist, box pleats are a popular style for skirts. Two pieces of cloth are folded in opposite directions and then brought together in the center of the fold to create a box pleat. A single box pleat is positioned in the back of a typical men's shirt for ease of wear (Fig. 7.19).

Inverted Pleats: In contrast to a box pleat, an inverted pleat is reversed. By folding the fabric inward, it creates a smooth effect. The fullness of the pleats helps create a smoother appearance by utilizing more fabric without adding bulk. An attractive addition to a skirt or pair of pants is an inverted pleat, which is a decorative element (Fig. 7.20).

Knife Pleats: This type of tight pleating, which is sometimes referred to as "knife pleats," "sunburst pleats," or "fan pleats," involves evenly folding the cloth all the way around. Accordion pleats, which create a zigzag pattern close to the are hem, frequently employed to give skirts more volume (Fig. 7.21).

Kick Pleats: A-line skirts look great with kick pleats because they give the garment's narrow shapes more ease and room to move. They give a garment a little fullness and are typically seen on the backside. They are reversed pleats. Kick pleats offer a tight

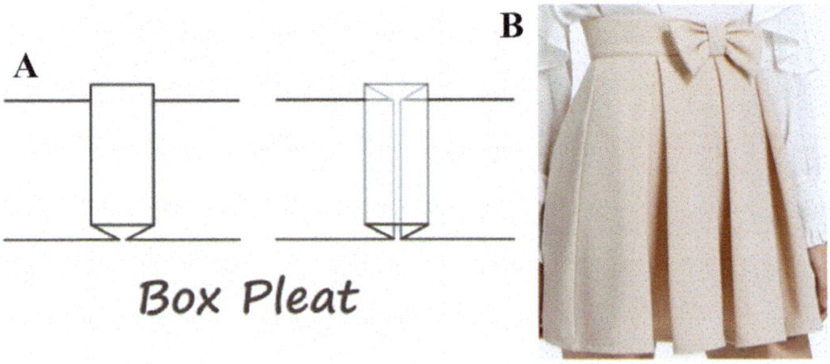

Fig. 7.19 Box pleats on skirt

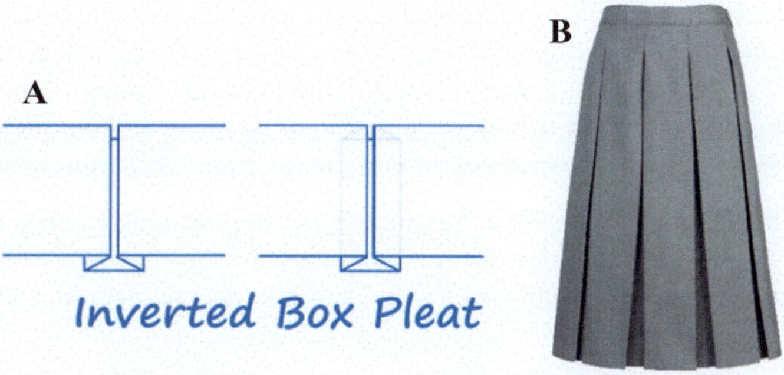

Fig. 7.20 Inverted pleats on skirt

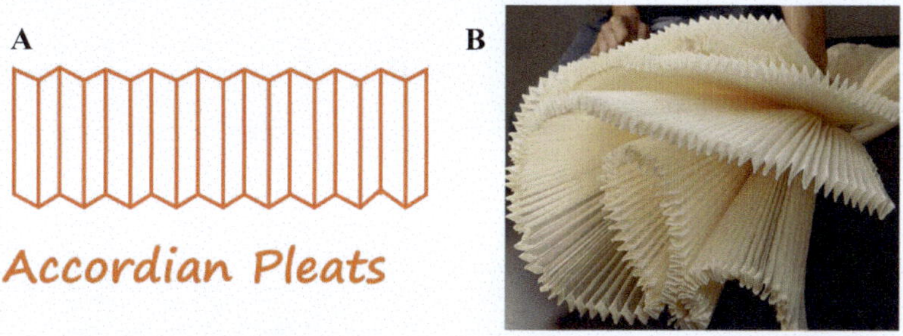

Fig. 7.21 Accordion pleats/knife pleats

garment a slight cut that allows it to breathe, giving it a "kick." They increase the user's mobility. Kick pleats are little pleats that usually appear in the back of skirts or jackets. They rise from the bottom hem of an item of clothing (Fig. 7.22).

Fig. 7.22 Kick pleat

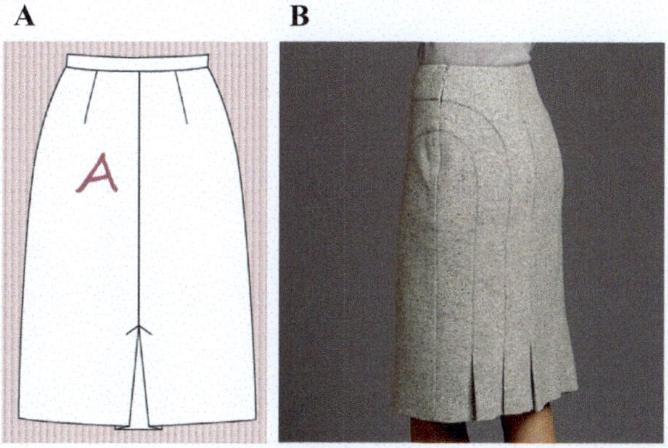

Forward Pleats: In dress pants and khakis, forward pleats are used to give the front of the legs more shape and dimension. Reverse pleats open toward the pockets, while forward pleats open toward the pants' zipper. These gained popularity during World War II, but because of a lack of fabric, pants had to be made without pleats, or with a flat front. For example, the majority of jeans today have a flat front (Fig. 7.23).

Honeycomb Pleats: A frequent technique for creating designs on pleated cloth is smocking, which is an embroidery technique that involves folding fabric into a pattern resembling honeycombs. Usually used as a decorative accent on cuffs and necklines (Fig. 7.24).

Organ Pleats: These pleats are frequently employed for skirts and to give the dress's bottom significant flayers. The equally spaced pleats around it give it the appearance of a pipe organ (Fig. 7.25).

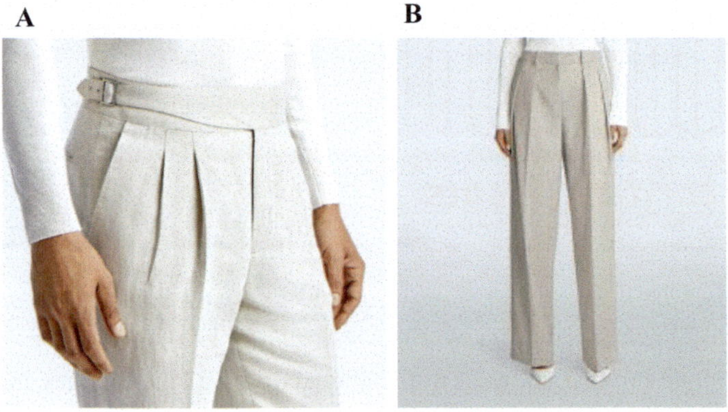

Fig. 7.23 Forward pleats on trousers

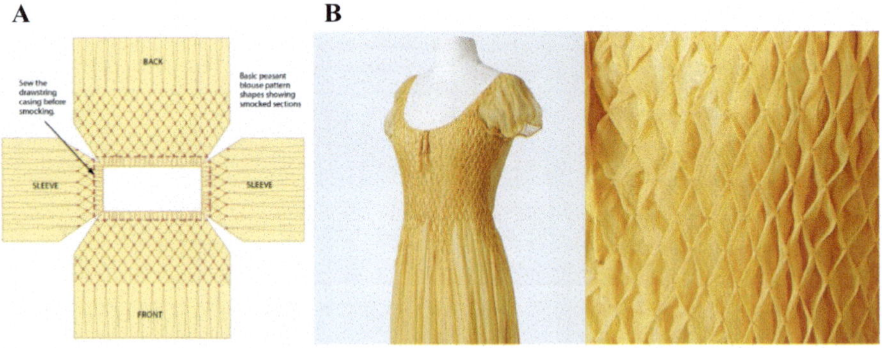

Fig. 7.24 Honeycomb pleats structure and dress

A B

Fig. 7.25 Organ pleats

Plissé Pleats: Like cartridge pleats, plissé pleats are retrieved, wet, and dried under pressure or underweight to create tight, flat pleats. The term "plissé," which is also related to "crinkle crêpe," originally described fabric that had been pleated or woven. The French word that means fold is the source of its name. This lightweight material now has a puckered, crinkled appearance that resembles ridges or stripes (Fig. 7.26).

Kingussie: These pleats were named after a Scottish town because they were so commonly used in kilts. The fundamental design of kingussie pleats is a box pleat, which is followed by knife pleats that enlarge on both sides, usually on the kilt's back. There are anywhere from 18 to 32 pleats in a traditional kilt (Fig. 7.27).

Cartridge Pleats: Sleeves and skirts are the most common garments to use cartridge pleats in because they create a sensual fabric without making the seam bulkier. Rather than folding, make two slack rows of stitches and gather the cloth by pulling the thread

A B

Fig. 7.26 Plissé pleats

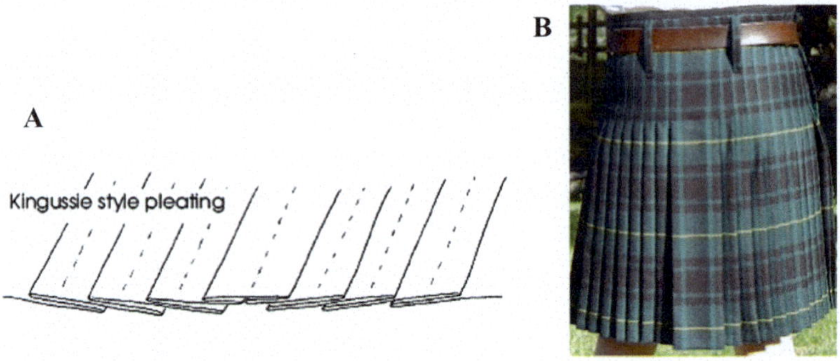

Fig. 7.27 Kingussie pleats

between them. It creates narrow pleats that are used to create bell-shaped skirts and puffed sleeves (Fig. 7.28).

Pleats in Skirts

- Adding fullness in waist, hips, and/or hem.
- For functionality such as kick pleats.
- As a design element; box pleats, knife pleats, and sunburst pleats.

Pleats in Bodice, Dresses, Sleeves, Blouses, Jackets

- For further ease and to allow the arms to extend upward and forward, box pleats can be placed in the center back of shirts or blouses.
- Soften the cuffs of garments.
- Using tuck pleats as a top and dress design element.
- Inverted box pleats are used in jackets and jacket linings to facilitate movement.

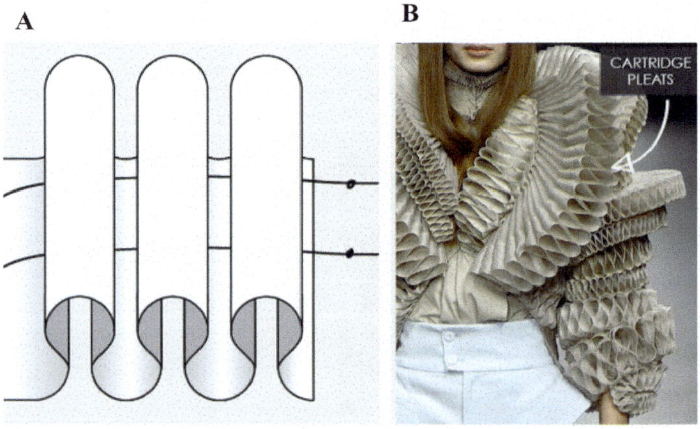

Fig. 7.28 Cartridge pleats

Hemming and Pressing Pleats

For all pleats, pressing is not necessary; however, if permanent pleats are needed, pressing is. The hem of a garment usually has pleats; pressing the pleat will keep it in place. Since pressing might be difficult to manage, various precautions are taken before pressing. To avoid press marks, press on both sides of the pleats, check that the folds are in the proper place, and slide paper under the pleat. It's important to leave extra seam allowance at the top when sewing pleats since more room is needed for folding and stitching the pleats. Pleats frequently call for a lot more fabric, so pay close attention to the fabric specifications to get the desired effect.

Pleats are generally utilized for both functional and decorative purposes in clothing and other textile applications. They may enhance the form, fit, and design of garments while providing visual interest and refinement. Pleats are versatile design elements that can be applied in a variety of ways to enhance the appearance, fit, and functionality of a garment. Each variety has a designated purpose, such as knife, box, inverted, and accordion pleats. With any luck, this post has improved your knowledge of the many pleat designs and how they can lengthen the figure, create the ideal appearance, or add movement. Once you know this, you can dress more stylishly by adding pleats to your clothes with confidence.

Tucks

Tucks, which are usually used for decorative purposes, are a great way to give any garment a high-end design element. Tucks are produced when a cloth fold is sewn straight. Think about them like stitched-together pleats. They are ironed after stitching and remain sewn horizontally in the intended direction.

Among the most often used tucks are the pin, dart, blind, and spaced tucks. These tucks come in a variety of widths and alignments and are quite simple. Since a tuck is represented by a single fold, the breadth of a tuck can be found by measuring the distance between the fold and the stitched line. Depending on how the garment is made, the tuck fold may be located inside or outside [13].

Basic Type of Tucks

Pin Tuck: Pin tucks are tiny folds of cloth that are stitched and pressed to one side. They are used on breathable materials used to make dresses, blouses, and other fitted clothes like coats, blazers, and slacks. Even though pin tucks are tiny and invisible, they dramatically change how a garment looks. While they have no effect on fitting, pin tucks may give the impression of a more slender, taller shape (Fig. 7.29).

Space Tuck: Wider tucks are those that are separated apart. The distances between the folds don't have to be equal; they might be uneven. Unlike blind tucks, spaced tucks don't always have to be placed the same distance apart from one another (Fig. 7.30).

Blind Tuck: The basic blind tuck is folded similarly to a pin tuck, but a standard blind tuck is usually larger and placed so that the folds of the individual tucks meet. In other words, if you look at the raw edge of a blind tucked cloth, as seen

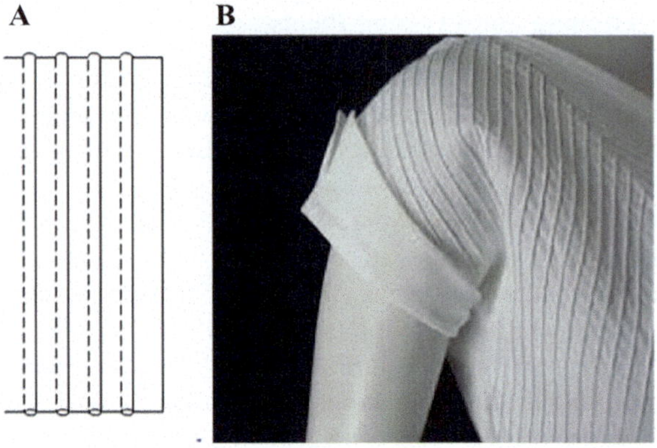

Fig. 7.29 Pin Tuck

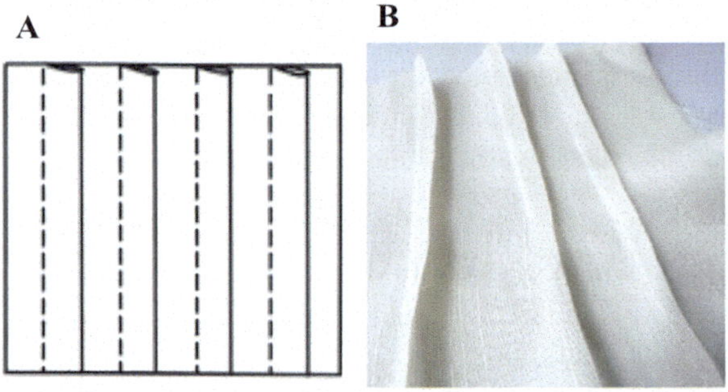

Fig. 7.30 Space Tuck

below, the next fold starts where the previous fold ended. This keeps blind tucks uniformly aligned and with the same amount of space between folds (Fig. 7.31). **Shell Tuck:** Typically, it is hand-stitched with tiny flowing stitches. The main purpose of this beautiful tuck is to complete baby girl's blouses and garments. Shell tucks give lightweight knits and woven textiles a dainty edge finish (Fig. 7.32).

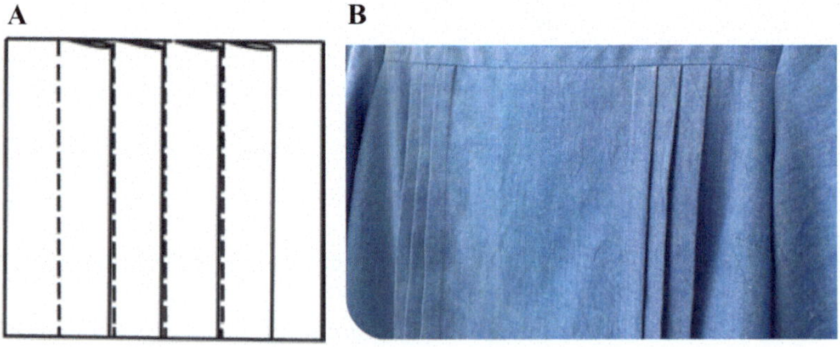

Fig. 7.31 Blind Tuck

Fig. 7.32 Shell Tuck

Dart Tuck: The dart tuck, also known as a double-pointed tuck, is a special tuck that's commonly used to control fullness by sewing it down and letting one point. The fit and form of a garment can be impacted by a dart tuck. Around the bust, waist, and hip area, dart tucks are a common feature of dresses and skirts. A dart tuck can be sewed and released on one or both sides, depending on the design you are working on. Additionally common is top stitching on dart tucks (Fig. 7.33).

Easing and Gathering

To join two unevenly trimmed edges together. There are two ways I can do this: by gathering and by easing. Relaxing controls, adding a little extra length, and collecting controls excessively.

Easing

A tiny amount of shape is achieved through easing. Relaxing the fabric in certain areas of clothes, such as the waistband, hems, set-in sleeve caps, and back shoulder seams, may be beneficial. When you're done, the relaxed area ought to be acne-free and smooth. It can be harder to relax some solid (non-stretchy) knit textiles than others. These materials include non-woven fabrics, woven fabrics with tight threads,

Fig. 7.33 Dart Tuck

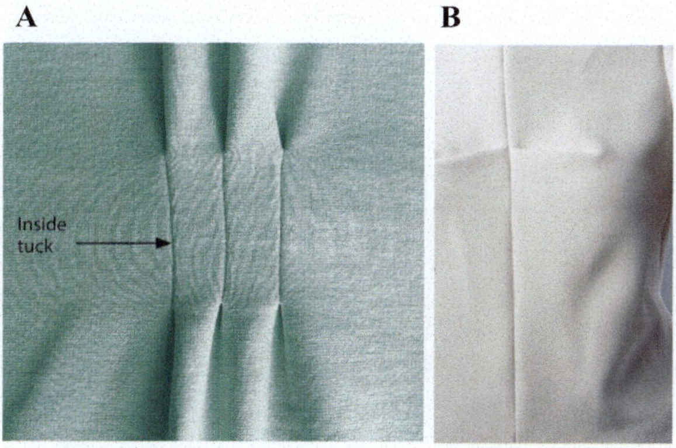

and fabrics with permanent press finishes. There are several situations where relaxing on certain pattern elements is necessary. Princess seams are often used when sewing a skirt to a waistline, flared hem, set-in sleeves, or at the shoulder seam.

Gather

A piece of cloth is collected by wrapping it to fit onto a shorter, precisely measured length of fabric in order to provide fullness. Gathering is usually achieved by using a longer machine stitch, which allows the threads to be tugged by hand, gathering the cloth up. The gathering is frequently done at one-third or half of the original width, depending on the fabric and the desired fullness level. The waistline, ruffles, sleeve cuffs, shoulder seams, and yokes are where the gathering is most commonly visible.

Gathers are used on the top of skirts to alter their size and add a subtle decorative element. This is accomplished by altering the general structure of the skirt by adding or removing soft folds without affecting its width. Gathers offer a smoother touch than pleats since they require less pressure to be applied and have a looser finish.

Methods Used for Fabric Gathering

Different methods are used to achieve fabric gathering. Some most used methods are given below with examples.

Hand Gathering: It is the most conventional method of assembling fabric. This technique is specifically suited for cloth that is thicker, more challenging, or ornamented and cannot be handled by a machine. Hand gathering can yield desirable and creative gathers (Fig. 7.34).

Machine Gathering: Machine gathering is the most common way of gathering it's more convenient because of its efficiency. It gives fine ruffles and skirt bottoms (Fig. 7.35).

Gathering with a Serger: It's a quick way to obtain gathers. It is applied on woven and thin, light fabrics. Sergeant must adapt to become gathered and difficult to manage (Fig. 7.36).

Cord Gathering: Cord gathers are perfect for tulle tutus and for hard fabrics that are difficult to handle (Fig. 7.37).

A B

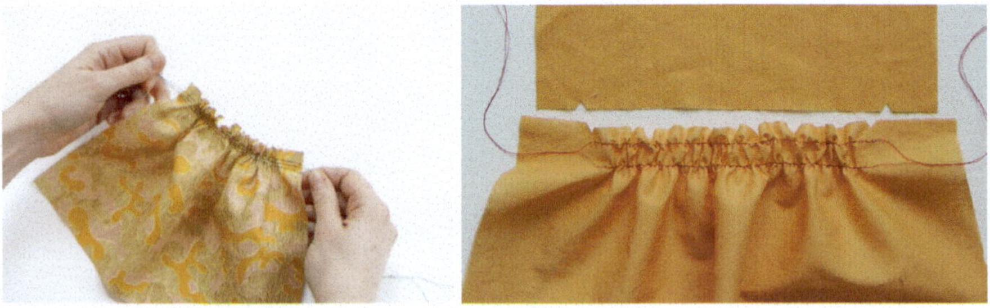

Fig. 7.34 Hand gathering

A B

Fig. 7.35 Gathering by machine

A B

Fig. 7.36 Gathering with a serger

B

A

Fig. 7.37 Gathering by cord

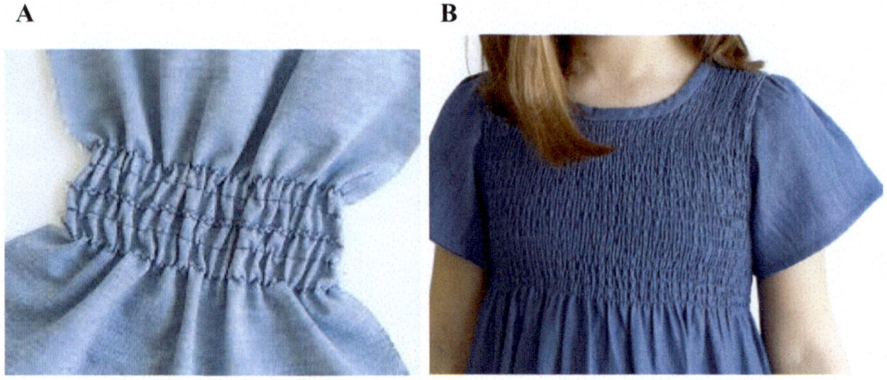

Fig. 7.38 Gathering with shirring elastic

Gathering with Shirring Elastic: Shirring elastic is primarily used on children's apparel to provide space and flexibility to clothing intended for expectant mothers. Elastic threads achieve the effect of shirring. Every clothing looks wonderful at a gathering like this (Fig. 7.38).

Conclusion

Based on the aforementioned studies, gathers, pleats, and darts are crucial components in pattern development and garment design. Designers are looking for fresh concepts, imaginative cutlines, unique fabric surfaces, and variations in dress fall that may be accomplished by utilizing all these many techniques as the fashion business develops. Above all, as knowledge grows, each person seeks to have a distinctive style that makes them stand out from the crowd. To this end, mastering the manipulation of darts, pleats, and gathers are essential skills for creating patterns.

References

1. R.H. Brown, M.L. Good, A. Prabhakar, *Apparel Manufacturing Glossary IIIOR Application Protocol Development*
2. Patternmakers' professional qualification and pattern making practices in Nairobi fashion houses, Kenya. A thesis submitted in partial fulfilment of the requirements for the award of the degree of master of science (Fashion design and marketing) in the school of applied human sciences of Kenyatta University (2017)
3. T.H.N. Nguyen, T.L. Nguyen, M.N. Nguyen, T.T.L. Tran, Research on the pleating technique by using cardboard pleat and Kraft paper. J. Tech. Educ. Sci. **72A**, 73–86 (2022). https://doi.org/10.54644/jte.72a.2022.1284
4. Y. Wang, C. Fowler, Y. Wu, *2018 Proceedings Cleveland, Ohio Basic Bodice Blocks Development for Fitting Large-Breasted Women*, 2018. [Online]. Available: http://itaaonline.org

5. REPORT RESUMCS ED 0!? 782 Objective of this course of study is to acquaint the student vii_ im The tools and symbols of patternmaking and to help him master the DASiC fundamentals of pattern development. II Follows the course of study approved by the board of education and was tested in various classrooms. theory and principles are combined with practical and creative application in costume design
6. Fulfilment of the requirements for a Doctor in philosophy in psychology (2002) [Online]. Available: http://eprints.mdx.ac.uk
7. A. Page, Creative pattern technology. Int. J. Fashion Des. Technol. Educ. **6**(2), 89–98 (2013). https://doi.org/10.1080/17543266.2013.793744
8. F. Berthouzoz, A. Garg, D.M. Kaufman, E. Grinspun, M. Agrawala, Parsing sewing patterns into 3D garments. ACM Trans. Graph **32**(4), 1–12 (2013). https://doi.org/10.1145/2461912.2461975

Chapter 8
One-Piece Dress Foundation

Fatima Iftikhar

Abstract In the realm of fashion, practicality often takes preference, and the one-piece dress stands as evidence of this principle. The versatility of one-piece dresses excels on occasions, making them ideal for a spectrum of activities. From professional settings requiring polished attire for work presentations to casual gatherings demanding a relaxed yet stylish ensemble, the one-piece dress effortlessly adapts. Its adaptability is a tailored asset, offering a rapid transition from day to night without the disturbance of wardrobe changes. Beyond convenience, one-piece dresses simplify the art of outfit creation, presenting a harmonious ensemble in one fell swoop. This not only saves valuable time but also minimizes the mental effort typically invested in mixing and matching separates. The ease of choosing a singular piece facilitates a polished look without sacrificing comfort, demonstrating that practicality can indeed reign supreme in the world of fashion.

Keywords Garment design · Fabric selection · Seam placement · Fitting techniques · Drape and structure

The One-Piece Dress: A Timeless Classic Beyond Trend

Introduction

The word one-piece garment is an umbrella term encompassing a wide range of garments and can therefore be defined by the following ways:

i. One-piece garment refers to production of a garment using a single fabric panel/piece and it does not require a seam/stitch to stay on body but may use fasteners and closures such as buttons or belt [1]. Examples are Skirts, Togas, Saris, Kimonos etc. (Fig. 8.1)

F. Iftikhar (✉)
Department of Textile Engineering, School of Engineering and Technology, National Textile University, Faisalabad, Pakistan
e-mail: fatima.iftikhar@ntu.edu.pk

M. Q. Khan et al. (eds.), *Garment Sizing and Pattern Making*, SDGs and Textiles,
https://doi.org/10.1007/978-981-97-7683-2_8

ii. Bifurcated garments covering both the upper and lower parts of the body. It forms a complete outfit in one piece of garment, and can require stitching and seaming to make it functional [1]. In theory, if it is a continuation as a whole even if it consists of two or more fabric panels, it is a one-piece garment. Examples are Leotards, Overalls, Dresses, Jumpsuits, Workmen suits, etc. (Fig. 8.2)

Fig. 8.1 An ankle length skirt. *Courtesy* Getty images

Fig. 8.2 Denim overalls. *Courtesy* Getty images

iii. In principle, a garment is established as "One-piece garment" if it doesn't require any other garment piece to classify as clothing. For instance, "Jumper skirt" and "Tunic skirt" are similar, but the former requires a top to classify as an outfit/clothing while the latter can stand on its own.

Historical and Cultural Significance

The evolution of one-piece dresses has woven a rich historical tapestry, reflecting the cultural identity and societal norms of different civilizations. In ancient Rome, the toga symbolized status and citizenship, serving as a unisex draped garment worn by both men and women. India's contribution to the world of one-piece dresses lies in the sari, a versatile draped garment that has stood the test of time. The sari transcends regional and social boundaries, showcasing an array of fabrics, colors, and draping styles that reflect the diversity of the Indian subcontinent. These garments were not just clothing; they were statements of identity, embodying cultural values and traditions. Across societies, one-piece dresses were worn by various social classes, acting as both a practical necessity and a symbol of societal roles. The evolution of these garments mirrored changing attitudes, reflecting shifts in power dynamics, cultural exchanges, and societal progress.

One-piece garment is not a novel idea. It has been a part of cultural and historical fashion in many parts of the world for a long time. Here are some examples of one-piece garments that have been part of their respective cultures for centuries.

Saari (India)

The Indian Saari involves the use of a long piece of garment that can be draped around the body in a variety of ways with each region having a distinct style. Some wore it with a blouse, while some covered their body entirely with just one piece of fabric. The garment requires minimal to no sewing and can be produced by using a variety of different fabric types. Due to its versatile, comfortable, and aesthetically pleasing design, it is gaining traction in the international market as well with Alexander McQueen basing their entire collection of Fall 2008 on the cultural garment [2].

Chiton (Greece)

A historical garment worn by both men and women from the Archaic to the Hellenistic period in Greece, the Chiton was a sleeveless shirt and was worn using linen or wool draped across the body in a variety of ways and then kept in place by brooches at the shoulders and a belt across the waist. Knee length or ankle length versions were popular among the women and the elderly, respectively while the men preferred longer versions of the garment (Figs. 8.3 and 8.4).

Fig. 8.3 A blouse-less sari [2]

Toga (Egypt)

In the early days of Rome, both men and women wore the toga, men wearing it without anything underneath except a loincloth; even later when it was worn nearly exclusively by men it continued to be worn by girls until they were 12. In its original form, it was a very handy and useful garment, which could even be worn into battle if you tied it up (Fig. 8.5).

Kimono (Japan)

A historical Japanese garment, still popular and in use today, is the Kimono. The garment is essentially composed of a flat body, sleeves and collar and is connected as one piece with no seams between the body and sleeves. The overall silhouette is a square line, with the line from the shoulder line to the sleeve being almost horizontal and then descending vertically toward the wrist. The trend of this one-piece garment continues into the twentieth century with prominent brands and designers such as Jeanne Lanvin, Balenciaga, Christian Dior, and Thierry Mugler reinterpreting the

Fig. 8.4 Chiton. *Courtesy* Getty images

Fig. 8.5 A woman wearing a toga. *Courtesy* Getty images

Fig. 8.6 Kimono from Kamakura era [3]

design in a variety of ways for their collections. A variety of different gusset designs, sleeve designs and wrap designs have been introduced [3] (Fig. 8.6).

Chemiloon (Manhattan, USA)

The chemiloon or the union suits were the predecessors of long johns and today's onesies and was patented in 1868. It was essentially a long one-piece under garment that emancipated women from the pinching confines of the corset. The garment combined a knit flannel shirt and pants into one piece. The long pants extended to the ankle, eliminating the need for stockings or garters. It also eliminated the use of heavy petticoats that often weighed around 15 pounds [4] (Fig. 8.7).

Origins and Early Forms

This section takes us on a historical tour, examining the one-piece dress in different eras. We'll explore how design and style have changed throughout the centuries. The roots of one-piece garments can be traced back to ancient civilizations, where early evidence reveals the ingenuity and practicality of human attire. In ancient Egypt, dating back to 3000 BC, draped linen garments were discovered, showcasing an early version of one-piece attire. These garments, often depicted in hieroglyphs and unearthed in archeological sites, featured simple yet effective construction techniques such as weaving and draping. The use of linen, a lightweight and breathable material, reflected both the practical needs and technological advancements of the time. Similarly, Mesopotamian artwork from 2300 BC provides glimpses of one-piece tunics. These garments, depicted in sculptures and paintings, showcase the early mastery of construction techniques like weaving. The use of materials such as wool

Fig. 8.7 Advertisement for promotion of Union suit featured in Harper's bazaar in 1879. *Courtesy* Stock montage/Getty images

and linen, as well as the incorporation of animal skins for colder climates, attests to the adaptability of one-piece garments across diverse environmental conditions [5].

Classification of One-Piece Dress

The one-piece garments can be classified based on the following categories:

1. Number of Sections
2. Fit
3. Garment type

Classification (Based on Number of Sections)

It can be classified into the following types based on the number of sections:

a. Bifurcated
b. Non-bifurcated

a. Bifurcated:

It comprises one-piece garments which cover both the upper and lower parts of the body, i.e., they have two sections such as jumpsuits and overalls [1, 6].

b. Non-bifurcated:

It comprises one-piece garments which only have one section such as robe, cloaks, etc. [1, 6] (Figs. 8.8 and 8.9).

Fig. 8.8 Denim jumpsuit. *Courtesy* Getty images

Fig. 8.9 A bath robe. *Courtesy* Getty images

Classification (Based on Fit)

It can be classified into the following types based on the fit:

(a) Loose-fit
(b) Make-shift
(c) Skin-tight

(a) **Loose-fit**

These one-piece garments do not cling to the body and often find applications in protective clothing [2, 3, 6, 7]. Examples include boiler suit, flight suit, coverall, etc.

Additionally, they are most likely to be traditional clothing specific to certain cultures. Examples include Saris, Toga, Kimono, Sarong, etc. (Fig. 8.10).

(b) **Make-shift**

Make-shift garments are improvised by using material available at the time of necessity, such as coming out of a shower. They are not considered fashionable but are functional and practical in certain situations [8]. Examples include:

Fig. 8.10 Saari. *Courtesy* Getty images

Fig. 8.11 A unitard. *Courtesy* Getty images

- Towel wrapped around the body for coverage
- Long t-shirt as a night gown etc.

(c) **Skin-tight**

These one-piece garments cling firmly to the body and closely follow the contours of the wearer's body. They are usually made out of stretchable material such as Lycra or spandex and are used in applications where long garments or an excess in fabric may cause inconvenience [9]. Examples include Cat suit, Leotards, Unitard, Swimsuit, etc. (Fig. 8.11).

Classification (Based on Garment Type)

One-piece garments can be classified on the basis of garment type as well. Following are the main categories:

(a) Under garments
(b) Standalone garments
(c) Outer wear
(a) **Under garments**

These one-piece garments are foundational garments worn under regular clothing [1, 9]. Examples include Body stocking, Corset, Shape wear, etc. (Fig. 8.12).

(b) **Standalone garments**

These are garments that do not require any other clothing article in order to be functional [2, 3, 6, 10]. Examples include Jumpsuits, Overalls, One-piece pajamas (Fig. 8.13).

Fig. 8.12 Corset. *Courtesy* Getty images

Fig. 8.13 Long johns/ one-piece pajamas. *Courtesy* Getty images

Fig. 8.14 Boiler suit. *Courtesy* Getty images

(c) **Outer wear**

These are one-piece garments worn outside of regular clothes and usually serve a specific purpose such as protection. Examples are One-piece cardigan, Workmen suit, Boiler suits, etc. (Fig. 8.14).

Fabric Choices: The Foundation of Form and Function

This section explores the relationship between design choices and the style and function of one-piece dresses.

Fabrics for Every Occasion

The choice of fabric plays a pivotal role in defining the comfort, aesthetic appeal, and suitability for various occasions when it comes to one-piece dresses. Here's an exploration of common fabric types and their distinct properties:

Cotton: Breathable and Casual Elegance

Cotton stands as a natural and breathable fabric, making it an ideal choice for casual dresses and warm weather. Its lightweight and absorbent nature offer comfort during hot temperatures, making it perfect for breezy summer dresses. Cotton dresses are not only practical but also exude a laid-back elegance, perfect for daytime outings or informal gatherings.

Silk: Luxurious Elegance for Formal Occasions

Silk, known for its luxurious feel and elegant drape, is a fabric of choice for formal one-piece dresses. Whether it's a sleek sheath dress or a flowing gown, silk adds a touch of opulence to the ensemble. Its natural sheen and ability to create flowing silhouettes make it perfect for formal occasions such as galas, weddings, or black-tie events.

Wool: Warmth and Structure for Cold Weather

Wool, with its warm and insulating properties, is a suitable fabric for colder weather. One-piece dresses made from wool provide both comfort and structure, making them ideal for wintery occasions. Whether it's a cozy sweater dress or a tailored woolen sheath, this fabric adds a touch of sophistication while keeping the wearer warm.

Knit Fabrics: Flexibility and Everyday Comfort

Knit fabrics offer flexibility and comfort, making them popular choices for casual one-piece dresses and everyday wear. From jersey to ribbed knits, these fabrics conform to the body's contours, providing ease of movement. Knit dresses are versatile and can seamlessly transition from casual outings to relaxed social events, offering a blend of comfort and style (Table 8.1).

One-Piece Production; Pattern Making Perspective

For one-piece garment production, one of the following two techniques is to be adopted.

1. Flat pattern production
2. Draping techniques

One-Piece Garment Production Using Flat Pattern Making

The one-piece garment production using flat pattern making constitutes production of garment via simple geometric shapes. This process is also catching traction for being sustainable by reducing the amount of cut and sew waste.

The production is explained with the help of a one-piece beachwear as follows [8].

Table 8.1 Examples of different styles of one-piece dress

Sr #	Garment	Definition
1.	Body stocking	Skin-tight, foundation garment (lingerie), covering the entire body except for head and hands [1]
2.	Body suit	Skin-tight garment covering the torso and crotch [9]
3.	Boiler suit/ coveralls	Protective, loose-fitting garment covering the whole body except for the head, hands and feet [1]
4.	Cat suit	Form-fitted outer wear garment covering the torso, the legs and frequently the arms [9]
5.	Crochet bikini	Swimsuit made out of crochet motifs [10]
6.	Infant body suit	Also called a Onesie, it is essentially a shirt which extends below the waist and consists of snaps that allow it to close at the crotch [5, 9]
7.	Jumpsuit	A one-piece garment with sleeves and legs and typically without integral coverings for feet, hands or head [6, 10]
8.	Leotard	A unisex skin-tight one-piece garment that covers the torso from the crotch to the shoulder [1]
9.	Jumpsuit	A loose-fitting casual Onesie for adults made of knit cotton (as used in sweatshirts), fleece, or chenille. They are mostly intended as lounge wear or sleepwear, but have gained significant popularity as stylish street wear [5, 9]
10.	One-piece ski suit	Sometimes called a "freedom suit". It covers the whole torso, arms and legs, and is worn while skiing [7]
11.	Snow mobile suit	A suit designed to be worn when riding a snowmobile. It is often similar to a one-piece snowsuit but specially made to not just insulate against snow and water but also to protect the rider from the wind while riding [7]
12.	Union Suit	A type of one-piece long underwear, most often associated with menswear in the late 19th and early twentieth centuries [4]
13.	Unitard	A skin-tight, one-piece garment with long legs and sometimes long sleeves, usually stopping at the wrists and ankles [9]
14.	Wrestling singlet	A one-piece, tight-fitting uniform, usually made of spandex/Lycra or nylon, used in wrestling. The uniform is tight-fitting so as not to get grasped accidentally by one's opponent, and allows the referee to see each wrestler's body clearly when awarding points or a pin [9]
15.	Zentai	A costume suit made of nylon-spandex blend that covers the entire body from head to toe [9]
16.	Kimono	A traditional Japanese garment characterized by long sleeves and a wide belt [3]
17.	Saari	A traditional Indian garment made of a single piece of fabric wrapped around the waist and over one shoulder. A versatile garment able to be draped in a variety of ways [2]
18.	Chemiloon	A one-piece long under wear that replaced corsets, petticoats and stockings as under garment for women in the 19th Century [4]

Production of One-Piece Beachwear Using Flat Pattern Production

A cover-up, which is essentially a lightweight, wrinkle resistant, practical, and versatile beachwear is used to bridge the gap between being fully dressed and wearing a swimsuit, can be made using flat pattern tools and techniques. A digital approach is taken to reduce waste and increase sustainable production.

Objectives:

1. To eliminate the need for changing into regular clothes from swimwear by providing an easy to wear, easy to carry, functional, and attractive alternative
2. Comfortable and light weight
3. Should provide multiple ways of wrapping around the body such as tying around the waist, draping it over the shoulders, or wrapping it across the body
4. Minimal waste production
5. Shaping the flat pattern to fit one or more body curves with minimal amount of stitching and seaming.

Requisites

Fabric Selection: Cotton, Linen, or Polyester chiffon can be selected, since all three are light weight and comfortable. Additionally, Linen is smooth in appearance while Polyester or Chiffon are wrinkle resistant. Moreover, knitted fabrics are preferred as they preserve their shape after cutting, eliminating the sewing and finishing additional steps, besides their ability to wrap and tie without the need for closures.

1. **Waste Reduction**: This can be achieved by minimal cuts and seam-free production. It also involves using the whole width of fabric.
2. **Ease and Style**: They include no closure and can thus be worn in multiple ways by tying or introduction of accessories such as brooches, belts, or magnets.

Pattern Production

The patterns are drawn in Adobe Photoshop in actual dimensions. The body measurements depend on the design.

Every one-piece pattern is focused on exploiting the whole width of fabric via the employment of geometric shapes that can be placed in the form of puzzle on the marker to increase fabric utilization. The diversity of shapes is to be considered. For some designs, the leftover fabric was cut straight for belt formation.

The knitted fabric edges can be cut and left without sewing as the looped structure prevents edges from tensile deformation. For the production under discussion, a knit-spandex blend is the popular choice due to the latter being a significant elastomer with excellent flexibility and wrinkle resistance.

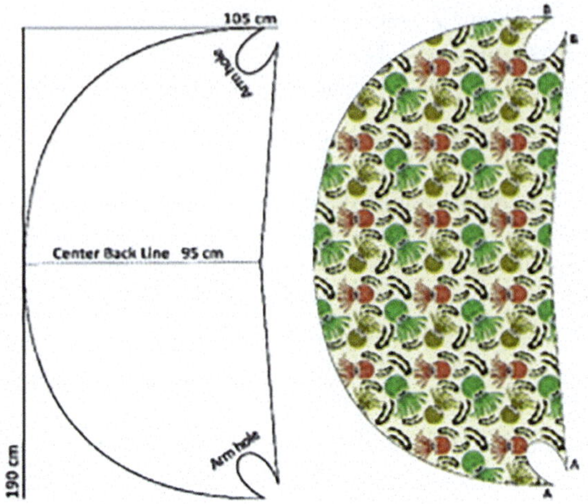

Fig. 8.15 Shoulder tied wrap [8]

Design 1: Shoulder-Tied Wrap

Backless, long V-neck cover-up wrapped around the body with straps formed by tying the fabric edges above the shoulders. The A-points are tied together on the left shoulder while the B-points are tied together on the right shoulder. The measurements of length, width, arm hole depth and distance of armhole to center back (Fig. 8.15).

Design 2: Halter/Sarong

It is the simplest design and requires only length and width. Waist circumference can also be used for varied designs (Fig. 8.16).

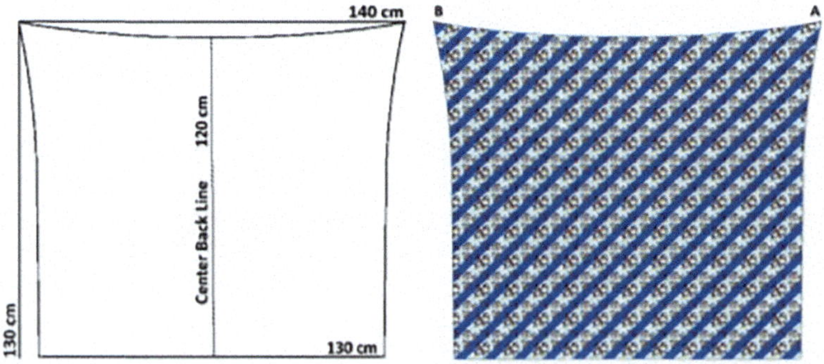

Fig. 8.16 Halter/Sarong [8]

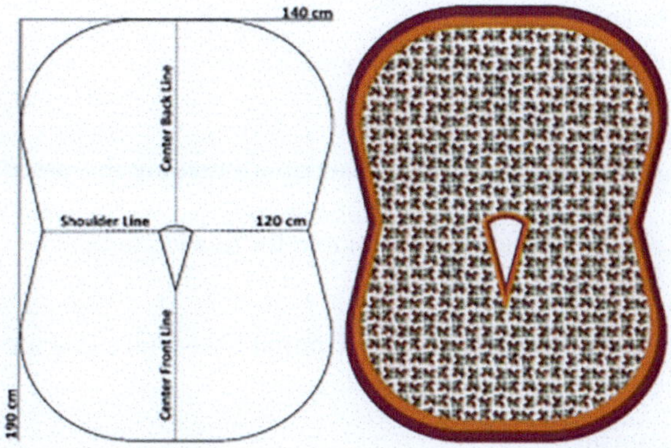

Fig. 8.17 Poncho/Kaftan [8]

Design 3: Poncho/Kaftan

A cover-up with a V-shaped neckline and a slit in the middle of the garment. The edges are curved symmetrically. The measurements of total length, total width, back width, and neck circumference are required for this design (Fig. 8.17).

Variations

Some design variations include:

- Side-tie skirt/One shoulder dress
- Asymmetrical tied skirt
- Cardigan with slits for arms

Other design components include:

- Combination of handicrafts with digital arts for creating contemporary textile printing designs such as the use of crochet motifs designed using CAD software or digital printing via sublimation transfer, that not only ensures higher production but also utilizes less water, dyes, and chemicals.
- Introduction of detachable or standalone accessories such as brooches, belts, or magnets.

One-Piece Garment Production Using Draping Techniques

Similar to conventional draping technique, this process involves the use of hanging and then pinning into place a test fabric/muslin to create fabric patterns which then serve as base for creating the actual garment [10].

Characteristics

1. Requires sewing
2. Produces fabric waste
3. Not as sustainable as flat pattern making discussed above
4. Better fitting garment production
5. Rigid fabrics are unsuitable for this technique

Pattern Production for a 4-panel One-Piece Garment

Requisites

The following material/equipment is required:

1. Dress form/Mannequin bearing correct measurements
2. Muslin fabric, similar to the fabric that will be used for the production of final garment
3. Four thread over lock machine for stitching the panels
4. Line tape/Style tape to indicate style lines
5. Steel pins to pin out the style tape as well as pinning and joining the various pieces of muslin fabric on dress form
6. Scissors for cutting excess fabric
7. Plastic Ruler of eighteen inches by 2 inches, marked with squared lines at 1/4th inch intervals
8. French curve for armhole and necklines
9. Hip curve for creation of hip curves in pattern
10. Sewing thread of good strength
11. L-Square with the long arm measuring 24 inches and the short arm 14 inches
12. Tracing wheel with small serrated edges to trace the pattern
13. Number 2 pencils
14. Large sheets of carbon tracing paper or white sheet for transfer of the lines from piece of the muslin to the paper

Pattern Production

The pattern production using draping techniques involves the following steps:

1. **Preparation of Mannequin**

A mannequin bearing correct measurements is essential for production of basic design by employment of Line tape draping. To keep the draping even across the body, the center line is marked with tape. Other lines of garment such as the neck line may also be marked with tape.

2. **Mannequin Measurement**

Mannequin is divided at the following points: Front Center Line, Back Center Line, Front Neck, Back Neck, Shoulder Line, Bust Line, Waist Line, Hip Line, Side Seam Line and Width across the shoulder, Front Chest, Arm Plate.

3. **Line Tape Draping**

The major points of line tape draping are Front Center Line, Back Center Line, Front Neck, Back Neck, Shoulder Line, Bust Line, Waist Line, Hip Line, Side Seam Line and Width across the shoulder, Front Chest, and Arm Plate. Black line tape is to be used for normal line tape draping while red line tape is to be used for princess lines i.e., lines used to produce final design.

4. **Preparation of Muslin**

The size of muslin required is to be estimated with reasonable room for ease, seam allowance, and styling. The edges are clipped. Tear with sufficient pressure is applied across the grain to break yarn evenly. The grain line is marked, the fabric is ironed, and seam allowance is added.

Four pieces of muslin for front and back, while one, each for collar and sleeve, can be used for basic pattern production.

5. **Draping Process**

The process can be divided into 6 parts:

(a) Draping the front
(b) Draping the side front
(c) Draping the back
(d) Draping the side back
(e) Draping the collar
(f) Draping the Sleeve

All of them involve the conventional method of draping wherein muslin is draped around the mannequin and fixed with steel pins as per line Tape and Princess Line.

Tucks and gathers are created and marked. The muslin is draped until the final shape of the garment takes shape on the dress form. Afterward, the fabric is removed and sent for truing.

6. **Truing and Tracing**

Pattern making tools such as rulers and curves are used to finalize the markings. The excess fabric is cut and removed and the fabric is again draped onto the dress form to test fit. After fit test, all seams and markings are recorded, and the final pattern is produced. The garment pieces are traced onto paper with the help of a tracing wheel.

(a) (b) (c)

Fig. 8.18 **a** Front view **b** back view **c** side view of the resultant garment. From left to right [10]

7. **Stitching**

Appropriate fabric is selected, and the muslin pattern or the traced pattern is placed on top of it. Seam allowance is added, and the fabric is cut. Ironing is optional. A sewing machine such as the 4 thread over lock machine is used for stitching the panels together.

The final result can be observed from the figure below (Fig. 8.18).

Sustainability and Ethical Practices: The One-Piece Dress Beyond Style

This section explores the growing importance of sustainability and ethical practices in the fashion industry, specifically in the production of one-piece dresses.

Sustainable Fashion

Sustainable fashion is an approach to clothing design, production, and consumption that prioritizes environmental and social responsibility. It seeks to minimize the negative impact of the fashion industry on the planet and its inhabitants, promoting ethical practices throughout the entire lifecycle of clothing. The core principles of sustainable fashion revolve around using eco-friendly materials, reducing waste, and minimizing the environmental and social footprint of the production process.

Cut and sew waste, accounts for a majority of textile waste. It is essentially the fabric from gaps and unused spaces between pattern pieces in a marker. A decrease in cut and sew waste won't only ensure sustainable production but also cost-effectiveness and higher efficiency.

One of the advantages of producing garment from a single pattern piece is sustainability by means of reduction of cutting and sewing waste. The patterns for production of such garments ensure the utilization of as much fabric area as possible in order to cut down on fabric waste.

Production of woven garments and knitted garments via the employment of one-piece patterns might provide the road where fashion meets sustainability. This can be achieved by minimal cuts and a seam-free production.

The geometric one-piece flat pattern created using digital techniques such as the employment of pattern making digital software, applications, and tools helps in decreasing textile waste by utilizing maximum amount of fabric to ensure less amount of cut and sew waste and introducing sustainability to the process.

The employment of knitted garment can help eliminate additional steps in sewing and finishing processes since they preserve their shape after cutting. Moreover, they are able to wrap and tie without the need for closures.

This, however, is not a recent idea. As mentioned before, the traditional one-piece garments such as Saari, Greek Chiton, or the Kimono are examples of one-piece garments that adhere to the zero-waste approach.

Combining handicrafts with digital arts for creating contemporary textile printing designs such as the use of crochet motifs designed using CAD software or digital printing via sublimation transfer, that not only ensures higher production but also utilizes less water, dyes and chemicals, is one other initiative to introduce sustainability in the process flow [8].

Conclusion: The Enduring Legacy of the One-Piece Dress

In conclusion, the one-piece dress stands as a strong force in the realm of fashion, weaving a narrative that excels times and cultures. From its ancient origins in draped linen to the iconic moments on red carpets and its adaptability to diverse lifestyles, the one-piece dress remains a symbol of timeless elegance and cultural significance. As we navigate toward a future guided by sustainability and ethical production, the one-piece dress has the potential to lead the charge, embracing eco-friendly materials and innovative designs. Its enduring appeal lies not only in its aesthetic allure but also in its ability to reflect and influence societal norms. As we bid ciao to this exploration of the one-piece dress, we recognize it not just as a garment but as a living testament to the ever-evolving interplay between fashion, identity, and the rich tapestry of human history.

References

1. U. Customs, B. Protection, What Every Member of the Trade Community Should Know About: Classification: Apparel Terminology under the HTSUS AN INFORMED COMPLIANCE PUBLICATION (2008). [Online]. Available: http://www.cbp.gov/
2. K. Kaur, A. Agrawal, Indian saree: a paradigm of global fashion influence. Int. J. Home Sci. **5**(2), 299–306 (2019). [Online]. Available: www.homesciencejournal.com
3. S.K. Kwon, S.K. Park, Study of design of kimono sleeve—focused on the design characteristics of the kimono that appeared indesign characteristics of the kimono that appeared in 20th-century fashion –. Res. J. Costume Cult. **23**(4), 595–603 (2015). https://doi.org/10.7741/rjcc.2015.23.4.595
4. M. Wrisley, Stella Blum Grant Report: 'Fashion I Despised': Charlotte Perkins Gilman and American Dress Reform, 1880–1920. Dress **33**(1), 97–110 (2006). https://doi.org/10.1179/036121106805253025
5. Woven Jumpsuits and One-Piece Garments, in *Complete Guide to Size Specification and Technical Design* (Bloomsbury Publishing Inc, 2016), pp. 177–194. https://doi.org/10.5040/9781501313059.ch-013
6. T.B. Petersen, Suiting children for institutions: the development, calibration and stabilization of the one-piece snowsuit. J. Des. Hist. **35**(1), 1–15 (2022). https://doi.org/10.1093/jdh/epab025
7. H. Abou Hashish, C. El Zean, One-piece flat pattern for printed beachwear cover-ups to minimize cut-and-sew waste and achieve sustainability. J. Des. Sci. Appl. Arts **4**(2), 206–221 (Jun. 2023). https://doi.org/10.21608/jdsaa.2023.191526.1253
8. S. Li, T. Ren, L. Zhan, A novel method for making a one-piece tight-fitting garment, in *2021 IEEE International Conference on Power Electronics, Computer Applications (ICPECA)* (IEEE, Jan. 2021), pp. 296–299. https://doi.org/10.1109/ICPECA51329.2021.9362580
9. Knit Jumpsuits and One-Piece Garments, in *Complete Guide to Size Specification and Technical Design* (Bloomsbury Publishing Inc, 2016), pp. 111–124. https://doi.org/10.5040/9781501313059.ch-008
10. W.H. Solangi, Development of one piece Fashion Garment with four panels using Draping techniques stitched with 505 stitch type. Marmara Univ. J. Sci. **27**(3), (2015). https://doi.org/10.7240/mufbed.86097

Chapter 9
Draping and Its Implication in Fashion Design

Gohar Ali Hayat, Rida Alvi, and Gangqianq Zheng

Abstract The covering is the earliest fundamental and physiological necessity for individuals from the beginning, and more aesthetically, it is to be dressed up to take the place of the covering. Clothing covers a wide range of physiological and psychological needs. Clothing is essential for all people's social interactions both physically and psychologically. That's why clothing designs are required to become unique, innovative, and responsive to people's aspirations and desires. Human beings demand clothing and variations in style and design. Draping is the technique of converting a two-dimensional garment design into a three-dimensional form. Draping goes back to 3500 BCE, beginning with the Mesopotamian and Ancient Egyptians. The organization of draped silhouettes like chiton, peplos, chlamys, and himation is popular in Greek fashion. All over history clothing has been classified as "fitted" or "draped". A "draped" garment like a toga doesn't need stitching, but a fitted garment would be sewn together and worn close to the body. The draping technique is used in the fashion industry to pattern both fitted and draped clothing. Designers prefer draping because it allows them the opportunity to observe their designs come into reality as they see the fabric in the dress form. However, a designer is first with a design sketch, during the draping process, a new and more fascinating design frequently emerges. Draping provides a designer with a sense of feel for the cloth as it drapes on the shape. Draping allows the designer to make better decisions about the appropriateness of a certain cloth for its design. Draping in fashion is the earliest way dating back to the eighteenth century. It is now considered a significant element of the design process. The layout of designing a garment is pinning the cloth on the standard size dress form. Different types and sizes of dress forms are available for ladies, men, and children to fulfill their demands through draping. A fashion designer can experiment with different types of fabric and how it falls to create new designs at the beginning of the dress design process. The toile fabric is separate from the dress that was used to create the sewing draped pattern for making stylish garments

G. A. Hayat · G. Zheng (✉)
Department of Industrial Design, Wuhan University of Technology, Wuhan, China
e-mail: 294350460@qq.com

R. Alvi
School of Art and Design, National Textile University, Faisalabad, Pakistan

M. Q. Khan et al. (eds.), *Garment Sizing and Pattern Making*, SDGs and Textiles,
https://doi.org/10.1007/978-981-97-7683-2_9

to fit an individual. Designers use a very lightweight fabric called 'tile' and designer mostly call it 'muslin' cloth.

Keywords Draping · Fashion design · Garment sizing · Pattern

Introduction

The prehistoric and most common type of clothing is draping. By draping, cutting, and pinning fabric on model form, the designer develops ideas and creates patterns for sample garments. Although most draping is done in muslin, the designer must keep in mind the properties of the fabric to be used for finishing the garment. The hand, weight, construction, and surface finish of the fabric all contribute to the final effect of the design. Some fabrics have unique properties and draping directly in the fabric of the finished garment is best. This however requires an experienced hand because the cost of most fabric makes mistakes prohibitively expensive [1].

Muslin is a pain weave fabric of unfinished cotton. The direction of the grain in muslin is easily visible, and its relatively low cost permits free use of experimentation and development. Muslin can be marked with pencil lines and the finished muslin pattern which is the end product of draping, can be used repeatedly. After the garment has been draped on the dress form, the muslin pattern, when stitched together, may also be used to adjust the fit of the human body. Once the designer has mastered the basic principles of draping, he or she is free to translate an endless variety of ideas into finished garments. Various types of fit can be developed to achieve the current fashion silhouette. Garments can fit close to the body, be relaxed and easy to allow freedom of movement, or be created in various degrees of oversizing. The loose fit of oversizing is often very fashionable. In addition, draping accommodates endless variations in body dimension and results in freely flowing apparel when soft fabric is used. Tailored garments are precisely oversized, and the fabric is reinforced with interfacing, backing, and various degrees of padding to achieve a well-defined silhouette. The body be relaxed, and easy to allow freedom of movement or be created in various degrees of oversizing. The loose fit of oversizing is often very fashionable. In addition, draping accommodates endless variations in body dimension and results in freely flowing apparel when soft fabric is used. Tailored garments are precisely oversized, and the fabric is reinforced with interfacing, backing, and various degrees of padding to achieve a well-defined silhouette [2].

Numerous Roman and Asian sculptures, terracotta, cave paintings, and wood carvings have been found that represent men and women wearing the same clothing while utilizing diverse wrapping and draping techniques. Some conspicuous examples of draped clothes from the Vedic era include Uttariya and Antariya. In the prehistoric era, Kasaya is a rectangular piece of Buddhist robe is the authentic illustration of the draped garment. Therefore, Sari, Odhni, and other more sophisticated draped fabrics are available. Draping, fabric, and body are about working together, but hard to master a skill and therefore is respected. Artistry, the minimum cut shape and refurbishment, body shape and beautify or basing. According to another definition, live or adaptable to the body with the help of a dummy (very thin and soft) with fabric or paper pattern using the method of preparation is the rehearsal techniques. With this method, evening dresses, more comfortable drape products are prepared in molds. According to Çağdaş draping, asymmetrical, draped, very artful models feature preparing pattern of clothes to see the finished shape, using paper and fabric clothes on mannequins is called to create an instance of the process.

Draping Tools:

Draping tools are mandatory to drape, measure, mark, and draft designs [3] (Fig. 9.1).

Muslin:

A low-cost fabric on which conspicuous grain and cross grains is used for draped woven goods-based fabrics.

Awl:

The sharp tool for drilling holes so that button and belt eyes can be installed.

L-square:

A ruler made of plastic or metal that has two arms of different lengths.

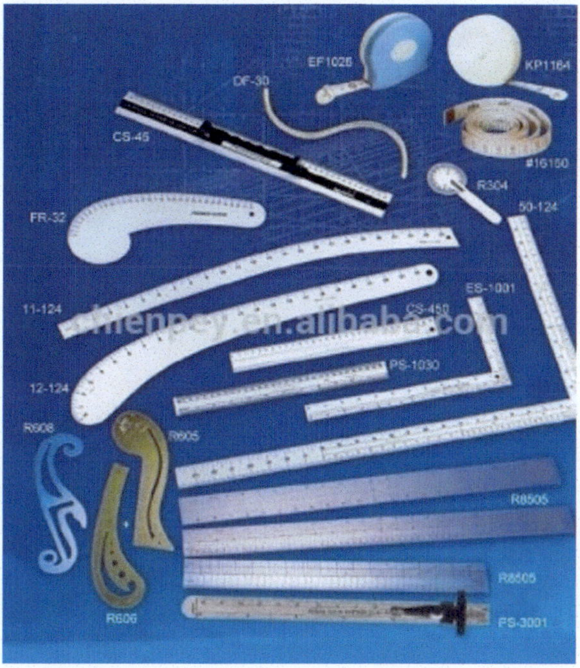

Fig. 9.1 Tools for draping

French Curve Ruler:

A ruler with an uneven curve that is used for curve and outline and edges of armholes, hips curve crotch seams, collars, and necklines.

Iron:

A steam and dry iron that is used to flatten, smoothen, and help block muslin.

Notcher:

A punching tool for marking the edges of a paper design sloper.

Plastic Ruler:

It is a two-inch extensive ruler with grids that are each 1/8 inches in wide.

Pencil:

The formation of muslin patterns that requires the use of a pencil.

Style Tape:

A thin, woven tape is used to outline fashion lines on a dress form.

Pin Cushion and Pin Dispenser:

A sewing tool that comfortably organizes pins places.

Scissors and Shears:

Shears are normally four to eight inches long and made of steel. Shears with bent handles are great for quick and precise cuttings.

Yardstick:

A ruler that is used for measuring hemlines which are made of metal or wooden one yard in length with marking in inches.

Tracing Wheel:

The wheel with a handle and a sharp, spiked edge, is specifically used to shift marks from the draped to pattern paper.

Tailor's Chalk:

A tiny piece of chalk with two tapering edges that is about 1/2 square that used to temporarily draw lines on numerous garments and other modifications point.

History and Background of Draping

The invention of drapery cannot be attributed to a single person. This method has developed over a long duration. Some works of drapery were found in ancient Greek and Egyptian designers.

Here are all the historical facts about draping that author found.

Ancient Egyptians and Mesopotamians that is cutting the fabric without seams and used loops tied up the pins they emulate the armholes and necklines. They used drapery to make attractive garments that were made from linen and other different textile materials. Even though the practice of drapery has evolved throughout the time, primitive method has not changed.

However, Greek and Roman fashion designers used the draping technique for designing their garments so that the ancient Egyptians were not the only ones who used these methods. The evolution of draped shapes like the chiton, peplos, chlamys, and himation guides to the expansion of Greek fashion. The Etruscans and Ancient Romans originated the toga it is a type of fabric that drapes and wraps around the body. Fashion designers use fabric to make robes and dresses that are draped elegantly over the body. In the fifteenth century, Flemish painter Rogier van der Weyden popularized painting portraits with cloth drapes in the background. A new type of painting called 'cloth painting' was born as a result [3].

In the sixteenth and seventeenth centuries, the use of drapery increased in the fashion industry. However, a trend toward more structured clothing in the eighteenth century caused draping to disappear. In the nineteenth century, draping came back when it was utilized to make magnificent ball dresses and theatrical costumes.

Draping has continued to be an important phase in the fashion design process. It is still used today to create elegant and sophisticated garments that gracefully hug the body.

Old Draped Dresses Design from Ancient Time [4]

Palla:

Palla is a long, rectangular piece of fabric that Roman ladies wore as a cloak. It is one of the most well-liked draped garments.

Stanapatta:

Stanapatta is a breast-covering chest band.

Peplos:

Long, draping clothing called peplos was worn by women in ancient Greece. It was frequently open on one side, had a deep fold at the top, and was secured across both shoulders.

Sari:

Traditional kilts and belted plaids are two examples of draped clothing from South Asia. Saris are often wrapped around the waist with one end draped over the shoulder, partially exposing the stomach.

Togas:

Romans covered themselves with togas, a very long piece of woolen cloth, leaving the right arm free while hanging it over the left arm and shoulder.

Antaraia:

A lower body garment called antaraia.

Angvastra:

An angavastra is a straightforward, loose garment that is thrown over the shoulders and frequently matched or paired with a dhoti.

Vashti:

In Tamil Nadu and the North and East of Sri Lanka, a veshti, also known as vi, is a white unstitched fabric wrap for the lower body.

Uttariya:

In ancient India, a garment referred to as an uttariya was worn by both men and women to cover their upper bodies. Unsewn fabric or a scarf was known as an uttariya.

Adivasah:

Adivasah is a form of overclothes from the Vedic era that is reminiscent of a mantle or cloak.

Why Draping is Important:

For the fashion students draping is the major technique in fashion design and technology. It permits them to visualize three-dimensional garments as a result instead of using flat pattern-making methods like cutting on the fold. Furthermore, it also gives students vital practice utilizing techniques for sewing darts, using facings, and working with bias-cut pieces during garment assembly, boosting their confidence for when they go on to flat pattern design when these abilities are once again needed. The techniques of draping are still implemented in high fashion, couture fashion houses, evening wear, and lingerie industries, where it is used to make a variety of garments. Before writing down anything on paper, the designer may quickly rectify any fit or design issues while draping a garment to see how her design creation will appear on the body. Additionally, certain design patterns simply cannot be produced with flat pattern making and must instead be draped. Some fabrics must be tested out on a dress form to determine how they respond [5].

- You may try out new forms, patterns, fashions, and materials when you drape. Without stitching any cloth together, you may quickly and easily test out new concepts.
- Draping enables you to design clothes that look great and fit the body correctly. A flawless fit that seems attractive and flattering may be achieved by adjusting the fabric around the body.
- Draping teaches you how to manipulate the fabric into the shapes you want.
- It aids in giving fabric the ideal form and silhouette.
- Draping teaches you how to work with cloth, a talent that every fashion designer has to have.
- Greater design versatility is possible with draped fabrics since they are simple to modify.
- Draping permits apparel designers and fashion designers to recognize and what creates a great fit and how to achieve it. The fundamental understanding may be applied to making any type of clothing's, including children's clothing, men's clothing and women's clothing. It applies to woven, knit, and stretch fabric as well, when necessary, pattern alterations must be performed.

Method and Process Used in Draping

How to drape

The way you drape will rely on your unique taste and preferences, thus there is no universal solution to this problem. All fashion designers need to be aware to assemble remarkable drapery techniques. Let' us take a look at them now!

The first step is to make a sketch or a reference design. As a result, the procedure will go more smoothly because you'll know where to start and can always improvise as you go. If you are feeling adventurous and want to jump right into it and go with the flow, that is probably fine as well.

The next essential stage in drapery is the choice of the appropriate fabric. Not all fabrics are created equal; some are better suited for voluminous draperies while others are better suited for more structural garments. Additionally, you will also want fabric's weight because heavier fabric would be more difficult to work with than lightweight fabrics. The suggested fabric is muslin because it is reasonably priced. After choosing a fabric, it is time to start the draping. How to drape is as follows [6].

Adjust the fabric to your specifications as you drape it over the torso or mannequin.

The next step is to design a fundamental pattern. This may be accomplished by folding the fabric in half and then cutting off a straightforward form, such as a square or rectangle.

When you have your initial pattern established, you can begin to add more specifics. To give your draperies greater shape and structure, for instance, you may use darts, tucks, and pleats.

The fabric should be pinned in place and cut by the design.

The last step is sewing the fabric together. Make sure you select a sturdy thread that can withstand deterioration. It could also be necessary to use a heavier-duty needle if you are dealing with a piece of heavyweight cloth.

And that's all there is to it! By following these simple steps, you too can learn how to drape like a pro.

Model or Mannequin

A professional model form is essential for draping professional garments. Various types of adjustable forms that are adequate for fitting are available for dressmaking purposes. However, they do not provide the solid, uninterrupted shape and surface quality required for pinning the fabric to the form, which is necessary to create the original garment pattern.

Standard model forms are used to design the dresses as well as separate tops and skirts. Pants are draped and fitted on bifurcated pants specifically designed for that purpose. The forms required for most apparel do not exactly replicate the human body. The undulations of the body are smoothed out to provide a simplified surface on which to work. Nevertheless, from are molded to the exact measurements specified for each size. Specially molded forms are also available for body-fitted apparel, such as swimsuits, brassieres, and bodywear, and specially sized model forms are available for coats and suits. Most designers of women's apparel work on a model form in the sample size used by their company. This form may be molded to the body measurements of the live model used for fitting or it may be a standard size. Many companies use a size 8 for samples.

The seams on the model form are precisely and conventionally placed. They are used as reference points in draping and must be properly aligned to result in well-balanced patterns. Occasionally, the seams on some model forms have been shifted, and their positions must be corrected.

Align shoulder and side seams so that the shoulder seam and side seam form a continuous line from the neckline to the waist passing through the center of the plate screw at the arm plate. Use the edge of the tape measure to determine the correct alignment.

- The side seam from the waistline to the lower edge of the torso should be a plumb line to the floor. Correct, if necessary.
- Check that the waistline tape is placed so that it flows in a smooth, continuous line around the narrowest point of the waistline.
- Make all corrections on the model form by sinking pins into the form at the corrected seam intersections.

Fig. 9.2 Muslin fabric

Selection of Fabric for Draping

We must take into consideration our active routine and way of life while choosing fabrics and clothing. The best fabrics for fashion draping are stretchy knitted fabrics, chiffon, georgette, organza, sheer cotton, voile, and lightweight silk. Drapability of fabric plays an important role in making sophisticated draped garments. For draped clothing, extremely light to medium-weight fabric works well. When the fabric is woven, it is typically draped in muslin or another cheap fabric that shows both the grain and cross grain well. Generally speaking, the muslin's quality and feel should mimic the properties and texture of the real fabric that was used to create the garment designs. Soft muslin is used to mimic the drape of natural or synthetic silk and knitted fabric. Mostly medium-weight muslin will mimic the drape of wool and medium-weight cotton fabrics. Heavyweight wool and cotton will mimic the raping quality of coarse muslin. When canvas muslin is taken into consideration, it will mimic the drapery features of heavier-weight materials like denim or faux fur. The less costly knit fabric should be used to produce the knitted clothing. However, it should be noted that the knit sample should have the same amount of stretch as the fabric chosen for the completed item. The front basic pattern or garment is often draped over the right side of the front dress form. This is the general guideline for fitting and draping any sort of clothing, but specific stages may differ. The way a cloth hangs or falls on a dress form or a human body is referred to as the draping method (Fig. 9.2).

Fashion Draping Techniques

Draping is the technique of wrapping fabric around a dress form to make a garment. Natural falls of fabric are used to achieve this. Uninformed draping uses a different technique such as pleating and gathering to create a dramatic effect. If you want

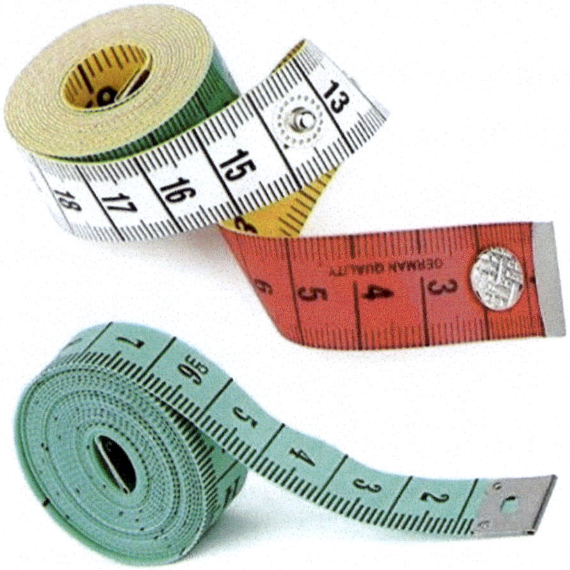

Fig. 9.3 Measuring tape for draping

to master this technique you need to understand every aspect of draping. There are several basic steps for draping to get you started [6]:

Prepping:

The initial step is to have the perfect and precise measurements of garments. The next step is to use tape to mark the dress's center line. This will keep the draping even across the dress (Fig. 9.3).

Sketch:

Drawing a rough sketch or illustration of your garment is usually helpful while designing a dress. It helps us to give an idea of how to go for draping. The sketch will help to understand how to effectively manipulate the fabric to get the desired dramatic and imaginative effects (Figs. 9.4 and 9.5).

Muslin:

The most lavish fabric is used in the draping process; it is flexible in its thickness and volume. Designers have a chance to experiment and play around to fulfill their requirements.

Tearing:

Estimate the size of muslin needed allowing a reasonable amount of extra fabric for ease, seam allowances, and styling.

Fig. 9.4 Basis sketch before Draping 2

Clip the edge of the muslin with scissors and tear across the grain with sufficient pressure to break the yarns evenly.

The true crosswise grain or the true lengthwise grain should form the edge of the muslin used for draping. Because the selvage is closely woven, it ends up holding in the muslin and often will throw the weave off; therefore, the center front and center back should be planned at least 3 inches from the selvage.

Blocking:

Muslin may need to be shaped before draping to get the yarns of the cross and longitudinal grains to be exactly at right angles to one another. Pull the edges of the muslin diagonally to reshape it into a square.

Seam Allowances:

The type of sewing machine used in the production of clothing has a significant impact on seam allowances. Any width of the seam can be sewn with a traditional lockstitch machine. The size of the seam allowance is determined by the type and placement of the seam, the potential need for alterations, and the final garment's retail price. Machines for over-edge and safety stitching, extensively used in knitwear, sportswear sleepwear, and children's wear require seam allowances ranging from 1/4 to 1/2 inch.

Fig. 9.5 Basis sketch before Draping 1

Beginning students of draping are usually expected to add 1-inch seam allowances on all straight seams, including side seams, waistline seams, and underarm seams. All curved seams are planned with 1/2-inch seam allowances.

Truing:

- Turning the muslin is drawing the lines that define the exact dimensions of the finished pattern.
- Remove the muslin from the dress form, and remove all pins except those indicating the vanishing points of the darts.
- Check that the dart cross mark at the waistline is equally distant from the apex grain line. Adjust if necessary.
- True the waistline dart by drawing lines from the vanishing points through the waistline cross mark to the lower edge of the muslin. Extend all lines for seams and dart beyond the dimensions of the pattern.
- True the shoulder dart by drawing a line from the apex to the crossmark nearest the neckline. Mark the vanishing point on this line at pin level; connect the shoulder crossmark nearest the armhole with the vanishing point.
- Draw the side seam by connecting the cross marks at the armhole and waistline.

Fig. 9.6 Pinning of draped design

- Lower the armhole for a basic set in the sleeve by measuring down on the dress form, from the armhole ridge at the shoulder over the center of the plate, to the side seam. At the side seam of the dress form, mark the depth of the lowered armhole with a pin. Determine the distance between this pin and the lower edge of the arm plate. Lower the armhole at the side seam of the muslin accordingly.
- True the armhole by placing the French curve so that the straight edge touches the shoulder seam and armhole cross mark

Pinning:

Pinning is a process that pulls the folds of the fabric together to create the draping design. By properly pinning the fabric, one may receive a visual image of how the finished garment will look (Fig. 9.6).

Trimming:

The final phase is shared by all design methodologies. Trim any extra fabric that may remain once the draping is complete. The raw edges of the drape should always be properly buried in the seams to ensure a perfect and clean appearance. Each of the above-mentioned techniques is designed to help make the process of draping easier and develop draping skills (Fig. 9.7).

Basting the Fabric:

After completing draping and pinning in this step baste the cloth. To finish the draping procedure, either stitch the fabric to the base piece or sew the fabric itself (Figs. 9.8 and 9.9).

Fig. 9.7 Trimming of extra fabric

Fig. 9.8 Final look of garment, front part

Fig. 9.9 Final look of garment, back part

Advantages of Draping in Fashion Design:

- It aids in the clear visualization of the garment. Draping allows fashion designers to gain a good notion and see how a garment will appear at the end even before they start cutting, stitching, and designing the costume.
- The most significant advantage of draping is that it enables for customization. A dress or a specific garment can be made to fit the person's dimensions and needs.
- The result is reduced fabric waste which you can try out before cutting the fabric.
- The draping technique also allows you to create various paper patterns and designs that may be utilized to perfect the next outfit.
- Because it is a three-dimensional technique, the design can be visualized while draping and any necessary change can also be made.
- By draping multiple fabrics, the designer can accomplish challenging patterns. For designing the cowls, this method can be used which gives good results.
- The designer can experiment with many designs and create new ones with the use of fabric falls to being the apparel garments productions.
- For a fashion designer, it is very useful because it is a time-saving technique.

Conclusion

This research led to the following conclusions regarding textile and fashion design education: Using the draping technique to recognize human anatomy and body structure helps prepare appropriate clothing, and using models and proofs without seeing the entire person is determined to help create custom clothing. It provides the opportunity to work as a sculptor of fabric draping the fabric designer fabrics that are hand can see the downfall of flexibility and can recognize the fabric. Prepare yourself for seeing a garment designer working with a dummy by helping you create a three-dimensional viewpoint and enhancing your creativity and imagination, which cannot be transferred to paper. To enhance demand for apparel, the foreign garment industry has long employed draping techniques. Draping is now seen as a crucial step since it enables fashion designers to experiment with fabric, use their imagination, and determine where darts, tucks, and other design elements fit better besides how the fabric behaves to the body. Mostly ladies preferred to wear draped garments. Fashion draping appears in movies, plays, and other media. Fashion technocrats use draping techniques to create evening dresses, stage wear, and party-wear clothing. The designers used this technique to create new and innovative designs with the fabric falls to begin the garment-making process. Additionally, a two-dimensional or three-dimensional draping name clothing design has emerged as a consequence of certain educational institutions' courses titled three-dimensional design draping garment patterns of course taking into consideration the lessons of the property. Increased training course hours for draping in textiles and fashion design are necessary, as is the provision of a more liberated and convenient working environment. To generate more unique shapes, clothing designs should be designed with various themes in mind. Prepared by draping premium audience overseas, taking into account that choosing clothing draping well-trained and will appeal to clients abroad should be trained in design. Draping is the best technique for the beginner student of fashion design to explore his/her skills from basic design to unique and critical design.

References

1. A. Forbes, J. Louise Clark, Development and Evaluation of Individualized Instructional Media for Draping on the Dress Form Item Type Text; Thesis-Reproduction (Electronic) (1978). [Online]. Available: http://hdl.handle.net/10150/555116
2. R. Gupta, Hybrid-Flipped class room approach for fashion design students: mitigating impacts to learning activities due to emergence of COVID-19, in *2020 11th International Conference on Computing, Communication and Networking Technologies (ICCCNT)* (IEEE, Jul. 2020), pp. 1–6. https://doi.org/10.1109/ICCCNT49239.2020.9225669.
3. R. Rana, S. Swale, *Article Draping As A Designing And Making Process, With Potential For The Unexpected* (2014)
4. Fashion clothing inspired from historical and traditional costumes via antibacterial treatment (2016). [Online]. Available: https://www.researchgate.net/publication/350956452
5. F. Ji, R. Li, Y. Qiu, Simulate the dynamic draping behavior of woven and knitted fabrics. J. Ind. Text. **35**(3), 201–215 (2006). https://doi.org/10.1177/1528083706055753

6. Y. Cho, T. Komatsu, S. Inui, M. Takatera, Y. Shimizu, H. Park, Individual pattern making using computerized draping method for clothing. Text. Res. J. **76**(8), 646–654 (2006). https://doi.org/10.1177/0040517506066966

Chapter 10
Digital Pattern-Making Techniques

Muhammad Tauseef Khawar, Ansa Al Kashifah Razzaq, and Hareem Fatima

Abstract Patternmaking is an essential component of apparel and fashion design. It plays an important role in translating the conceptual design of an individual into existence. Patterns help the designers to ensure the accuracy of each component of design element to fulfill the requirement of the customer. Effective patternmaking paves the way toward economical, consistent, and well-fitted apparel production. Traditional patternmaking involves manual pattern drafting, draping, and flat patternmaking methodologies. It also requires more time and resources to perform design work. However, technological advancement and digitalization played a significant role in patternmaking. Nowadays, digital and advanced methods are widely being used to make the 2D and 3D apparel patterns with accuracy and consistency. In the garment industry, various types of software are being used to create patterns, manage designs, and simplify the production process. Software-based tools help designers, patternmakers, and manufacturers in creating accurate patterns, grading, and fit evaluation. In this chapter, we have described the software-based advanced technologies that are currently being used in garment industries and have made work easier, efficient, and accurate.

Keywords Patternmaking · Design elements · Conceptual design · Technological advancement · Digitalization

Introduction

The art of patternmaking flourished in the fifteenth century, transitioning from uncut fabric to carefully tailored pieces. From the eighteenth century, garment fitting evolved from manual adjustments to systematic drafting methods. The invention of the standard tape measure revolutionized tailoring, leading to sophisticated drafting systems. In the nineteenth century, there was a significant push to accelerate

M. T. Khawar (✉) · A. Al Kashifah Razzaq · H. Fatima
Department of Clothing, School of Engineering and Technology, National Textile University Faisalabad, Faisalabad, Pakistan
e-mail: tauseefkhawar@ntu.edu.pk

M. Q. Khan et al. (eds.), *Garment Sizing and Pattern Making*, SDGs and Textiles,
https://doi.org/10.1007/978-981-97-7683-2_10

garment production to satisfy the growing needs for affordable clothing driven by the expanding middle class. The pattern drafting systems of the nineteenth century formed the foundation for ready-to-wear sizing, and with the rise of mass production, interest in individually fitted patterns declined. Home economists, following East's model, then primarily focused on drafting custom patterns [1–3].

At the onset of the twentieth century, a significant shift occurred in the tailoring industry, increased public interest in fashionable clothing led to the organization of tailors working together under the same roof of organized batch production utilized the individual skills of tailors, who still managed the entire garment-making process but employed available drafting methods. Consequently, clothing manufacturing transitioned from a cottage industry to industrial-scale production. By the conclusion of World War II, shifting market requirements were evident, at the same time emergence of liberated women favored simpler, more comfortable clothing styles, prompting a transition from menswear to womenswear production. The Industrial Revolution highlighted the necessity of standardized patterns for the success of ready-to-wear fashion, leading to the development of systematic size charts like those used in the US and UK. In bespoke tailoring, measurements were traditionally marked directly on the fabric using tailor's chalk for each customer, a method also employed in home sewing known as tailoring pattern cutting [4, 5].

A garment pattern serves as a stencil for cutting fabric into pieces that are assembled to create a whole garment, and garment patternmaking, also called garment construction design or pattern cutting, involves drawing garment construction lines on brown paper based on fashion sketches provided by designers. Patternmakers cut the paper along these lines to obtain paper pieces for fabric cutting, while this process acts as a connection between fashion design and clothing production and is regarded as a highly technical tasks within clothing companies, requiring extensive experience with free patternmakers. Two methods exist for making garment patterns: the traditional hands-on approach and the computer-aided process, both demand advanced skills, and mastery of these techniques typically requires several years, rendering current patternmaking methods [6]. Creative pattern cutting is both a design process and a technological art form but crafting a pattern demands expertise and understanding [7].

Historically, fashion's evolution has intertwined with scientific and industrial progress, sparking new technologies in the twenty-first century with advancements like 3D garment printing and virtual fabrication and innovative fashion forms are resurfacing. While CAD/CAM streamlines pattern drafting, fashion has been hesitant to adopt computational methods for creative design and craft-based approaches [8]. Fashion companies often face inefficiencies due to the separate processes of fashion sketch drawing and patternmaking, requiring repeated communication between designers and patternmakers. Various technologies are used to streamline these processes, such as 3D-to-2D flattening, virtual try-on, and rapid fashion sketch generation. However, previous research has not combined these two tasks, and the proposed method integrates fashion sketch drawing and patternmaking, ensuring that the generation of a garment pattern closely follows the fashion sketch but now the approach significantly improves the efficiency of garment product development [9].

Modern CAD systems offer a range of analysis tools to help designers optimize and test their designs' functionality [10]. These systems efficiently generate manufacturing information from CAD designs, while garment manufacturing technology has developed incrementally, design systems have focused on visualizing garments and efficiently creating design images, particularly for large retail organizations. These systems enable quick experimentation with colors and textures and can generate prototype textile prints using advanced printing technology. Additionally, there is a high level of automation in subsequent manufacturing processes, including nesting 2D patterns and generating cutter head path information for large cutting tables [7]. The apparel industry faces environmental challenges, with textile production contributing to carbon emissions and fabric waste. Zero-waste pattern design, led by designers like Rissanen and McQuillan, aims for sustainability and innovation. Computer technology, particularly CAD/CAM applications, has revolutionized garment design, streamlining efficiency, accuracy, and creativity [11].

Patternmaking

Pattern making is a strategic plan for the garment on the basis of which the fabric is cut and sewn, and hence, the garment is manufactured. It is the technical drawing or drafting of a garment and the most initial step in any garment manufacturing process. Garments are constructed using standard size charts, body forms, or figures to create 2D patterns [9]. Before cutting and assembling any garment, an overlay of the pattern is created usually on the paper according to the design of the garment as shown in the picture below (Fig. 10.1).

Patternmaking has become necessary which enables us to make different garments. Patternmaking plays a vital role to create designs clearly and understand it with technical ability [9]. Patternmaking is a critical process in the garment industry, serving as the foundation for producing well-fitted and aesthetically pleasing apparel. Flat patternmaking, drafting/drawing, and draping are the three basic methods of patternmaking [10].

Flat patternmaking is method in which a basic fitted pattern is created with adequate ease margin to fit a person. A sloper is the starting point in a flat pattern method as it serves as a basic block or foundation pattern that is made on cardboard or hard paper without seam allowances. Flat patternmaking is the fastest and the most efficient method to create any pattern design. Using slash, we can manipulate flat patternmaking, as it is a fundamental method for creating design patterns. To get a good pattern, the most important thing is accurate measurements. Without precise measurements, we cannot make a perfect pattern. The image below shows how flat patterns are formed using manual method [10] (Fig. 10.2).

Drafting is another method in which basic pattern is constructed by taking measurements directly from any dress or the person. This process is mostly used by home tailors or haute couture. The draft is then cut from muslin fabric for individual, with adjustments marked on the paper. Creating patterns in this method is

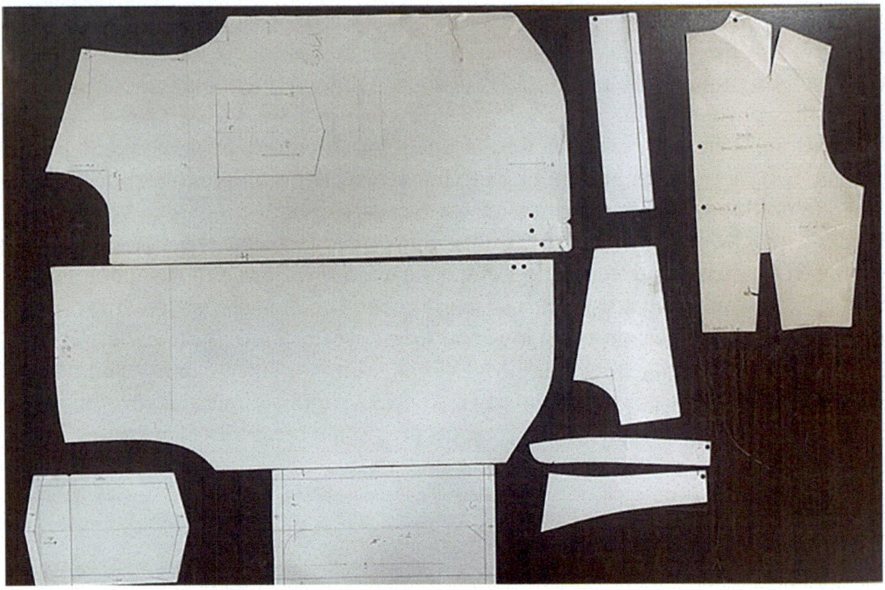

Fig. 10.1 Pattern is sketched on a paper before cutting the fabric

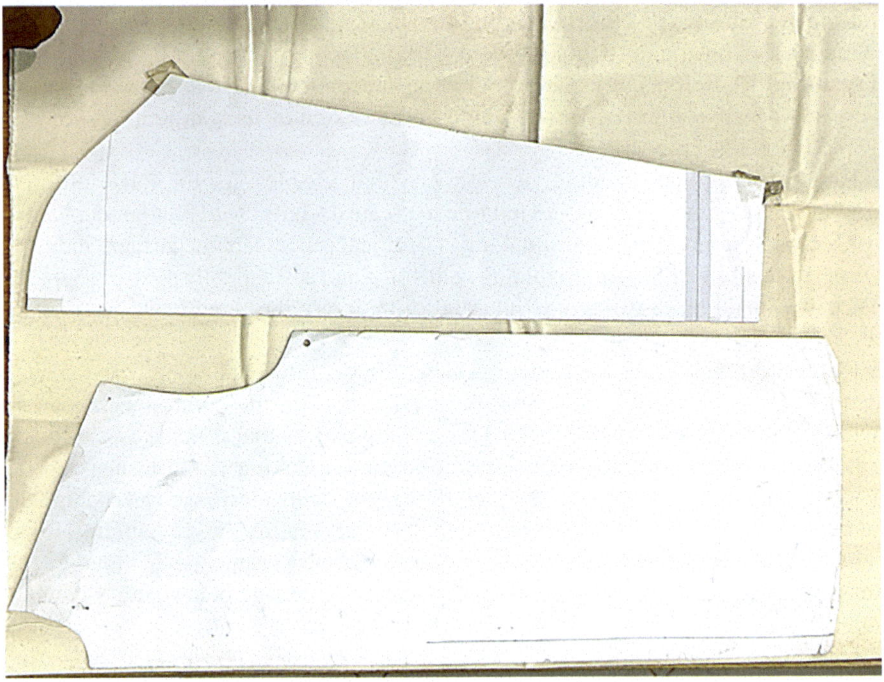

Fig. 10.2 Flat pattern is being made using sloper

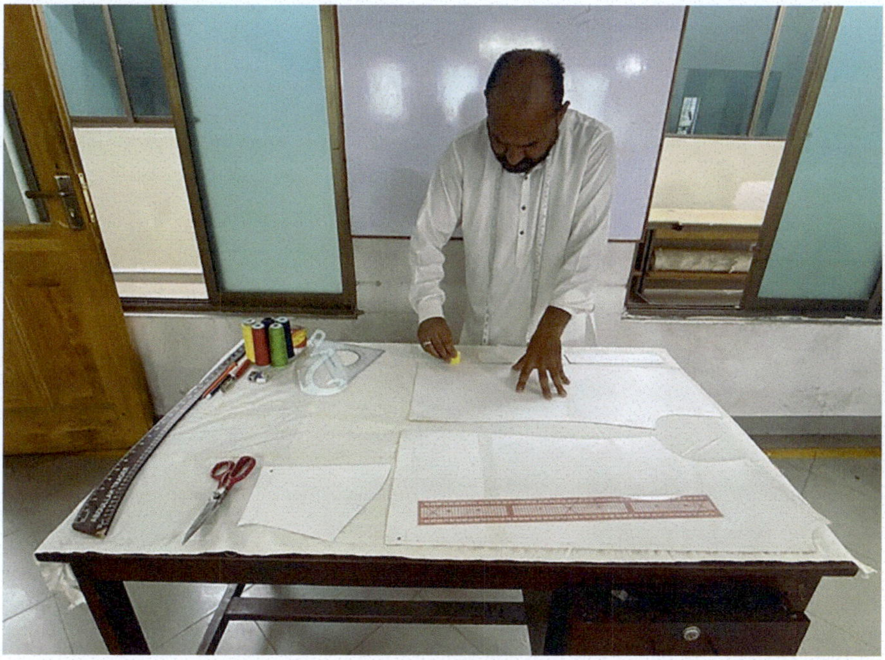

Fig. 10.3 Transfer of measurements from dress to the paper to draft pattern

straightforward as they are based on templates, but crafting the templates manually can be challenging due to the need for precise measurements and numerous test fittings. Following image shows the pattern drafting process [10] (Fig. 10.3).

Draping is a method in which a 2D fabric is wrapped around a 3D form (mannequin) to create a pattern of the garment. After creating a pattern of the garment on the fabric using mannequin, it is transferred to the paper. The fabric that is used to do draping is called muslin. The best thing about draping is we can see the overall look of the final garment before cutting and sewn on original fabric. This process is much time taking and expensive as compared to flat patternmaking. As shown in Fig. 10.4, you can see how draping is done using a 3D body called mannequin to achieve the final look of the required garment [10].

There are many different methods to make commercial patterns, in which not all of these methods include three-dimensional dress forms. In fact, patternmakers are now using advanced computers or even high-end tablets to make patterns that can be used to cut and assemble any kind and design of clothes just in a couple of times. No matter which pattern-making techniques are used, the process of pattern development begins when sketch of a garment is made in an initial draft. While this basic sketch is made manually in mind. After drafting, there are a few steps required to transform this initial draft to create a perfect fit garment [11].

Fig. 10.4 Drapped manniquin with different pattern styles

Advanced Pattern-Making Techniques

The apparel design is not limited to making a few sketch lines on paper for pattern drafting but involves a technical process to translate a novel approach into existence, which has evolved over a period. In the current era, digitalization played an important role in the advancement of pattern-making process. Nowadays mostly garment industries are using digital and advanced methods to make the patterns of required garment in 3D ways. In the garment industry, various types of software are used to create patterns, manage designs, and simplify the production process. These software tools help designers, patternmakers, and manufacturers in creating accurate patterns, grading sizes, and visualizing designs [12]. In the following sections, we have described the software-based advanced technologies that are currently being used in garment industries and have made work easier, efficient, and accurate.

Computer-Aided Design (CAD) Software

CAD software is used for creating detailed and precise digital designs of garments. Designers use CAD tools to draw and visualize clothing designs, experiment with colors and textures and create virtual prototypes [12]. Computer-aided design (CAD), also known as computer-aided design and drafting (CADD) in its full form, is a specialized technology for design and technical records specifically during the design stages of the manufacturing process. It is an additional program that replaces manual drafting with an automized process for creating 2D and 3D concept visualizations, essentially as a blueprint before the design is brought to life. CAD systems were originally developed in the early 1960s to design precision machines for construction and other complicated tasks but were adapted in the 1970s to suit the textile industry as well. Presently, CAD systems are used widely in both textile and fashion design globally [13].

CAD systems are used by engineers, architects, designers, and fashion managers as an easy way to create, alter, and enhance the design process overall. Design drafts can be quickly and smoothly scanned onto a computer and edited within a software program, rather than being hand-drawn many times and scanned only once a final design concept has been reached (as was the design process before the development of CAD technology). Designs uploaded to CAD systems can usually be stored in a cloud and referenced at any time for future use. CAD systems and technologies are highly advantageous because they enable the design process to be cooperative, more communicative, and more productive in the long term [13].

Although CAD and computer-aided manufacture (CAM) systems both greatly increase productivity and benefit the design and creation of manufactured textiles, many fashion designers and textile producers begin the production process by making a series of manual sketches by hand. Before the creation of CAD and CAM software, designers had to create the sketches by hand until the point of perfection. After designing by hand, the image is scanned and transferred to the computer. Then after scanning the design is drafted, and pattern is adjusted according to the sizes and design. Various CAD technologies are serving different purposes, such as software created significantly for reshaping the design of woven fabrics and software planned for printed items. One of characteristics of CAD system is illustration or sketch pad application that allows the designer to draw and experiment with different styles, while some software can be used to forecast different effects including print, drape, and silhouette in final product. Many CAD technologies are used to save various embroidery designs and prints for further usage anywhere [13].

To create a detailed design, for example like animal print, then the design is roughly drawn on paper. After scanning that hand-drawn design in a computer, CAD system is used to change the color of the print, change their geometric structure, change the color of the background of the shirt, and modify or rotate the print. Before transferring the final design to a CAM system, we can edit or reuse that design in our desired manner as many times as we need, as shown in Figs. 10.5 and 10.6 [13].

Adobe Illustrator

Illustrator is widely used for creating vector-based or scalable designs and patterns due to its flexibility and precise detailing [13]. Adobe illustrator is used for creating fashion drawings, technical sketches, textile and pattern designs, and presentation

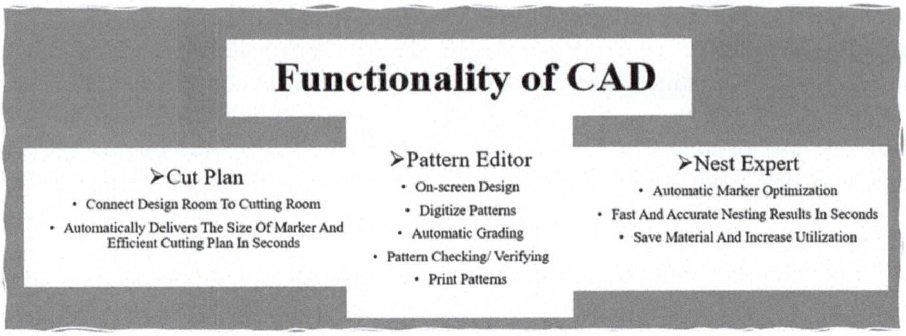

Fig. 10.5 Functionality of CAD software

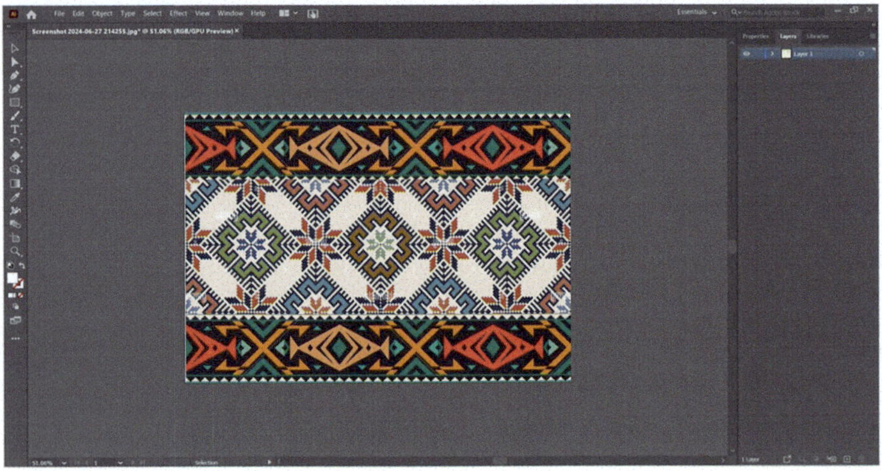

Fig. 10.6 Creation of presentation boards with different colors and patterns using illustrator

boards. 2D drawing of a garment, including styling details and construction details is called a fashion drawing. Garments in the factories are created according to the technical flats. Due to its versatile pattern tools, pattern repetition and customization has become very easy. Complex patterns in textile industries are also created by using different tools of this software. Illustrators also tend to create fashion presentation boards, which include elements like comprehensive themes, color palettes, fabric selections, images, and fonts. These boards serve as a mood board for industry. Before production every design is sketched out by pencil on a paper by designers to depict design details of the fashion product. But now, the latest technology is used for the majority of designing and sketching [14].

In the fashion sector, designers use this software to create simple to complex designs and patterns of any fashion garment or accessory by adding beautiful details like dots, dashes, shapes, or any other type of design. Original patterns are created by using pattern tool which can be repeated naturally throughout the design. After implementation, changes can be made in any aspect of design including size, color, dimensions, etc. In the fashion industry, presentation boards play a very vital part in creating the environment for different meetings and catalogue presentations. They aesthetically depict the idea of trends through a cohesive blend of theme, color scheme, fabric choices, styles representative images, and typography using fonts suitable for the relevant theme. Illustrator is very important tool for fashion designers in creating these theme boards which help them to plan, sketch, and customize different options efficiently with the relevant artboard tool, simplifying the process of design for comprehensive and meaningful presentations [14].

Adobe illustrator is a very powerful tool for creating detailed vector patterns, making it adaptable for designing prints, sketching fashion figures, and producing expandable high-definition images. Its patternmaking tools allow the customization of textures and patterns for precise technical drawings, and illustrator helps in specifying measurements, seams, and stitching techniques. Photoshop enhances the process with its digital prototyping qualities, enabling image presentation on virtual models. Moreover, textile manipulation, texture experimentation, and fabric finish simulation can also be done on photoshop by using its advanced tools. To convey design concepts effectively, cohesive mood boards, color adjusting, and refining the images are also done by designers using photoshop tools. To enhance overall design presentations with realistic shadows, highlights, and visually appealing collages, the advanced tools of software serve as invaluable gem [15].

CorelDRAW

Similar to illustrator, corelDRAW is distinguished as a well-known vector graphics software used to create patterns and illustrations [12]. It is known as the market's most widely used graphic design programs and serves to fulfill daily requirements of professional designers. This software is significantly designed to edit 2-D graphics like posters and logos, etc. Making it advantageous for business, its adaptably in creating both simple and complex shapes and lines for drawings clipart, logos, textile prints and different applications [16] (Fig. 10.7).

Due to its adaptable design properties, the apparel industry uses this software extensively for pattern designing. Designers can create complex layouts, templates, and patterns ensuring precise measurements vital for proper size and fit in clothing design. The object-oriented tools allow the creation of detailed designs in different formats and then can be altered and transferred to different file formats which enhances the compatibility with other tools in garment manufacturing industries. Due to its faster workflow, it aids in efficient file handling, image extraction, and object selection. Moreover, advanced tools including smear tools, twirl, attract, and repel tool offer creative possibilities for refining vector objects [16].

CorelDRAW excels in representing fabric and color combination for garments for creation of quick and flexible design color schemes based on specific needs.

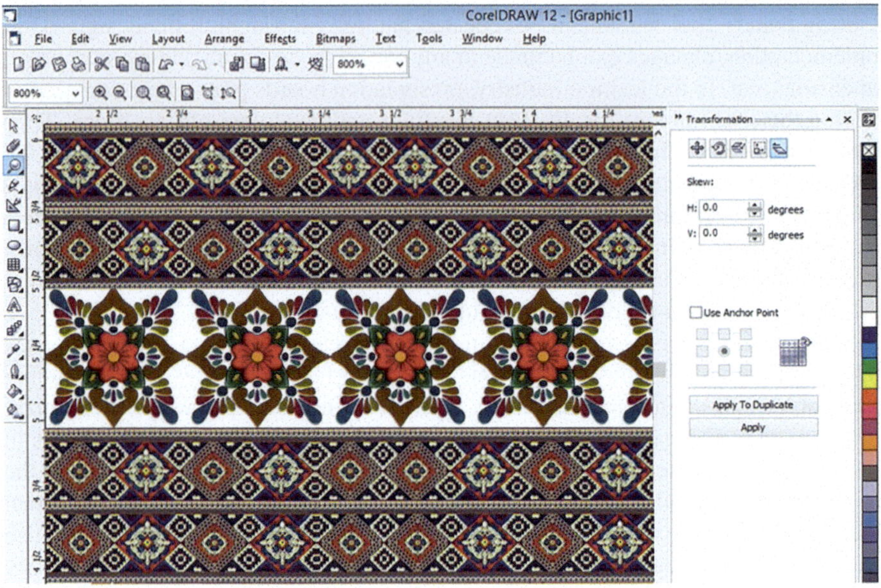

Fig. 10.7 Textile design formation using corelDRAW software

Its powerful drawing operations, which include the capability to draw unrestricted curves, changes in line styles, set drawing origins and add supporting lines, meet the non-standard nature of garment scaling. Due to its adaptability features, it is a comprehensive tool for digital clothing design in drawing settings, unit and ratio settings, and interchangeability of different file formats. By creating clothing effect charts, it represents the desired clothing effects very precisely [17]. The technical analysis of CoralDRAW software from six different aspects is shown in Table 10.1 [17].

The personality design in clothing refers to the stylistic design of garments, representing unique, creative thinking and artistic expression of the designer. It aids in better understanding of modern fashion life. During the whole creative process, analyzing the logical structure of the clothing style design is the most important initial step [17].

Optitex

This software is renowned for its 2D and 3D capabilities, empowering designers to create, visualize, and adjust patterns. With its cutting-edge pattern design software, Optitex enhances the production of goods to better meet the customer needs and to achieve the goal of quicker cycles. This software uniquely enlightens the needs of technical textile and garment industries by seamlessly integrating robust 2D design

Table 10.1 Skill-based technical analyses of corelDRAW [17]

Skills	Skill analysis
Basic line drawing	Edit the basic prototype with lines, then transform the interactive deformation tool with a zipper, and finally push and pull the deformation to continuously edit the lines
Line processing at the detail	The length, size, left and right, opening degree, shoulder width and sleeve link of the front sleeve, and the placement angle of the sleeve need to be determined. The interaction of the line tool and the guideline is very important.
Lines are combined into a confined space	To apply a closed path, fill elements are drawn in the target area.
Combination line	Combine different elements, such as symbols, lines, and images, into any individual.
Adjustment of line proportion	Drawing style plans, craft drawings, printed embroidery patterns, and a few people are also used to draw style renderings.
Line symmetry	By modifying the curve, length, neckline, sleeves, patterns, and colors, the original clothes are designed to be dragged to form a variety of different clothing styles.

with realistic 3D visualization on a single platform. By providing the display of finished garment's appearance, fabric behavior, and body movement, this simulated fashion show aids manufacturers to transform 2D designs into dynamic 3D images on virtual models. Prominent apparel customers of Optitex include Land's End, Christian Dior, Cherokee, Target, Morgan, Timberlake, Disney, Vanity Fair, and Guess Jeans [18].

Beyond the garment industry, its innovative animation technology has gained interest in gaming and movie industries, with Hollywood studios investigating its possibilities. Ran Machtinger CEO, founded in 1988, this software provides a range of products from the Optitex PDS (pattern design system) with high-end pattern design and grading tools to the leading 3D Runway Designer. This simulated 'try it on' system simulates worn garments on 3D mannequins based on user-defined parameters, allowing virtual garment production using both flat patterns and 3D technology. The 3D runway creator makes draping patterns on virtual models possible and can change the body measurements according to the need to evaluate the fit of item. Users can also create personalized size mannequins by simplifying the fitting process due to its 40 adjustable body measurements on 3D mannequin [18].

The new animation technology has expanded business into a new sector: the entertainment industry, specifically in designing the dresses of animated characters for computer games and movies to achieve realistic clothing. While the entertainment industry is using animated technology for years, Optitex technology is emphasized for its simplicity, speed, and efficiency compared to existing movie animation techniques [18] (Fig. 10.8).

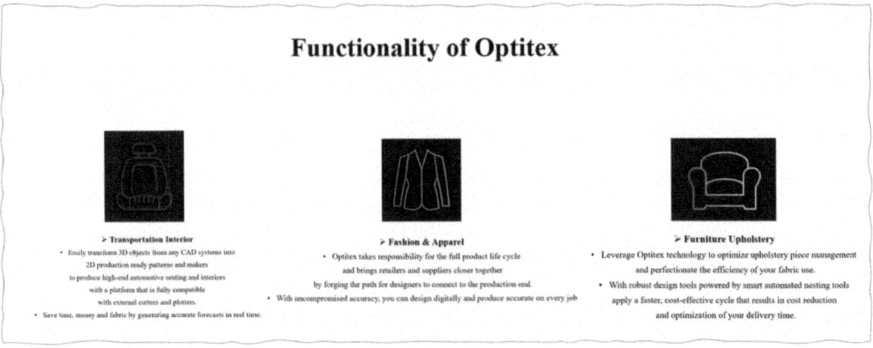

Fig. 10.8 Advanced functionality of Optitex in multiple domains

Gerber Technology

Gerber technology, a global provider of integrated software and automation solutions for the garment and industrial sectors, specializes in delivering technological solutions throughout various stages of the product lifecycle across diverse industries. Particularly renowned in the apparel manufacturing sector, Gerber technology offers advanced hardware and software solutions for automated processes such as cutting, grading, pattern creation, and marker creation. The traditional manual approach to creating patterns in different sizes has been significantly replaced by Garment Gerber Technology (GGT). By utilizing GGT, a single base size is all that is needed, enabling the swift generation of numerous designs through a process known as grading. Ambition, in its operations, utilizes state-of-the-art Gerber technology, with the support of plotters and digitizers ensuring precision and accuracy in every design they produce [19].

Gerber AccuMark transforms the design, development, and marker-making process by offering a high-tech suite of software applications, including Gerber AccuMark 2D, Gerber AccuMark 3D, and Gerber AccuNest. This comprehensive solution enables users to visualize designs, create precise patterns, and boost markers, all while seamlessly sharing data across the supply chain. Gerber AccuMark empowers fashion brands to expedite launch time, improves fit of the garment, and improves material utilization. Manufacturers benefit from AccuMark's marker-making technique to produce efficient fabric-cutting markers. Renowned fashion brands globally rely on Gerber AccuMark 2D for their design, development, grading, and marker-making requirements. With multiple grading options, automated processes, and innovative image functionality, Gerber AccuMark 2D ensures enhanced efficiency, increased yield, and improved communication, bringing creative design concepts to fruition in no time [20] (Fig. 10.9).

Gerber AccuMark 3D simplifies the design and development workflow by establishing a direct link between 2D patterns and 3D models. This integration allows

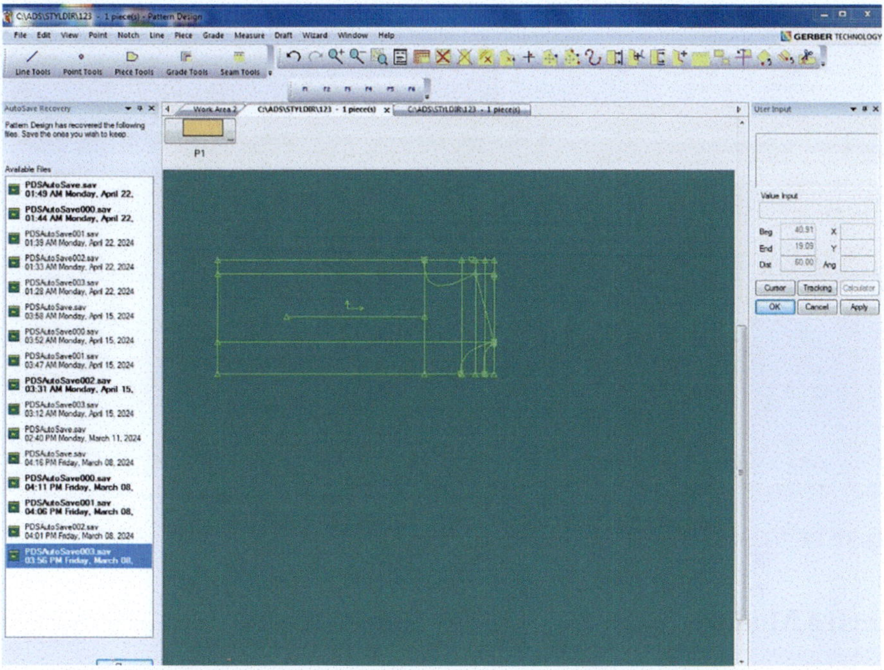

Fig. 10.9 Gerber AccuMark 2D for creation of 2D pattern of garment

users to enhance their concepts and refine fit through finalized virtual samples, ultimately minimizing the necessity for creating numerous physical prototypes. Gerber AccuNest is an automized nesting application crafted to save users time and money by creating cost and production markers spontaneously. Supported by robust algorithms, Gerber AccuNest remarkably speeds up the nesting process, improving material utilization and efficiently lowering costs [20] (Fig. 10.10).

Gerber AccuMark Made-to-Measure provides sophisticated automized features, covering pattern modification, advanced procedure-based specifications, and order creation. Streamline the process of generating patterns with measurements from various sources and automatically produce markers, enhancing overall productivity. Efficiency in the CAD-to-cut-room process is essential for true operational efficiency. Gerber AccuPlan facilitates this streamlining by enabling the optimization of cutting processes. It achieves this by generating advanced spread and cut plans, considering fabric rolls and cutting table availability. Additionally, AccuPlan allows the planning of multiple cut work orders concurrently, supported by a robust production tracker. The system further ensures seamless tracking of orders from the CAD department to the cutting room using sophisticated digital cut tickets and barcodes [20].

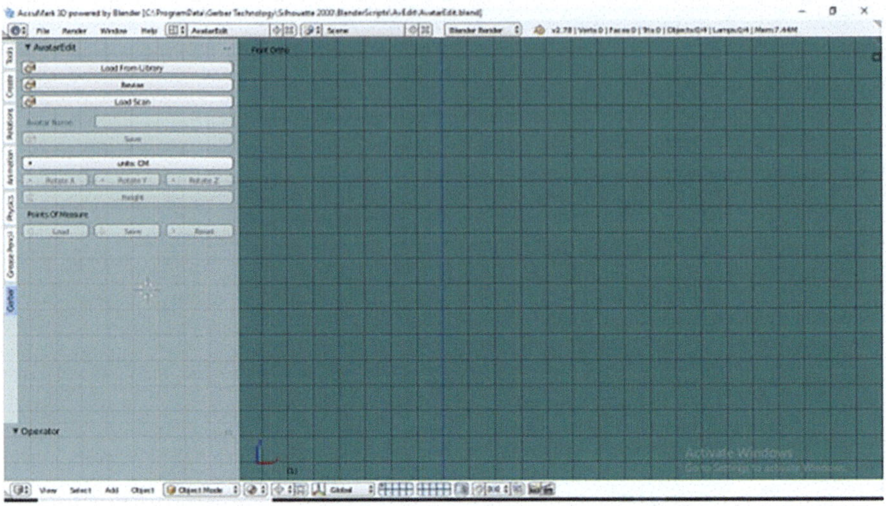

Fig. 10.10 Interface of Gerber AccuMark 3D

Lectra Modaris

Modaris, an automated pattern creation software widely employed in the garment manufacturing industry, is developed by Lectra. Fashion brands leverage Modaris to craft patterns and prototypes primed for production. Following industry best practices in product development, Modaris ensures a consistent standard of excellence and quality throughout the entire process, guaranteeing a flawless fit for clothing in all sizes. Notably, it allows simultaneous alterations to both 3D and 2D models. The capabilities of Modaris 3D include specialized tools for checking pattern fit, access to an extensive collection of textiles categorized by properties, and various other features. Modaris Classic is utilized for creating, modifying, grading, and industrializing patterns. To expedite the process, graded basis patterns can be generated, stored in a library, and subsequently employed for future models. Modaris Essential facilitates the grading and industrialization of designs stored in native Modaris files, enabling a swift and error-free sharing of pattern files—making it particularly advantageous for suppliers, especially during pre-production stages. This functionality enhances collaboration with brands, allowing for faster and more effective cooperation [21].

Modaris, developed by Lectra, serves as patternmaking software empowering fashion companies to generate patterns and production-ready prototypes. The software integrates best practices in product development, ensuring a sustained level of quality and excellence throughout the entire product cycle. Modaris expedites the product development process by enabling patternmakers to leverage their expertise, streamline pattern creation, eliminate repetitive tasks, and save up to 30% of pattern adjustment time, thanks to its robust parametric capabilities. In comparison with Modaris Classic, Modaris Expert offers an even higher level of efficiency. It allows simultaneous adjustments to be applied to both 3D and 2D models, ensuring a perfect fit for garments of all sizes. Modaris 3D introduces unique tools for verifying pattern fit, an extensive fabric library categorized by characteristics, and other valuable features to enhance the pattern creation and adjustment process [21].

Modaris Classic offers a comprehensive set of functionalities, allowing users to create, modify, grade, and industrialize patterns. The ability to generate and store graded base patterns in a library facilitates expeditious use for future models, contributing to process acceleration. On the other hand, Modaris Essential is specifically designed to grade and industrialize patterns stored in native Modaris files. Particularly advantageous for suppliers, especially in pre-production stages, this application enhances collaboration with brands by enabling the seamless sharing of pattern files, devoid of data loss or errors. A notable feature of Modaris Essential is its capability to automatically estimate fabric consumption, providing valuable insights for informed decision-making during the product development process [21].

Grading Software

The act of transforming a sample size, also known as base size, into a second, smaller, or bigger size is known as pattern grading. To grade patterns, one uses a size specification sheet. Grading modifies the size of the garment's original form rather than creating a new one, as shown in both of the following figures [22] (Fig. 10.11).

The primary objective of pattern grading is to proportionally adjust a pattern's size while maintaining the fit and shape of a garment. The origins of grading as a scientific process trace back to the nineteenth century, driven by the advent of mass manufacturing and the introduction of commercial patterns. To cater adequately to various body shapes, a design needs to be graded into a range of sizes. In contemporary practices, patternmakers and garment manufacturers typically begin with a middle-sized pattern (commonly a size S or M) and then grade it up or down to create larger and smaller sizes. There are three fundamental methods of pattern grading, with the first approach being considered the simplest. In this method, to grade up or down, the pattern is sliced, and the pieces are distributed by a specified amount. The essential tools for this process include a pencil, ruler, scissors, and measuring tape [22].

In the second method, patterns are moved using this manner at a fixed distance from one another. To get the same products as the earlier cut and spread procedure, the designer redrew the shape. The image above shows how the measurement of a pattern is shifted in using this method. The third one is computer grading that has advanced in parallel with the development of new technology. It is thought to be the quickest approach. It essentially digitizes the "cut and spread" and "pattern shifting" approaches using the same procedure [22] (Fig. 10.12).

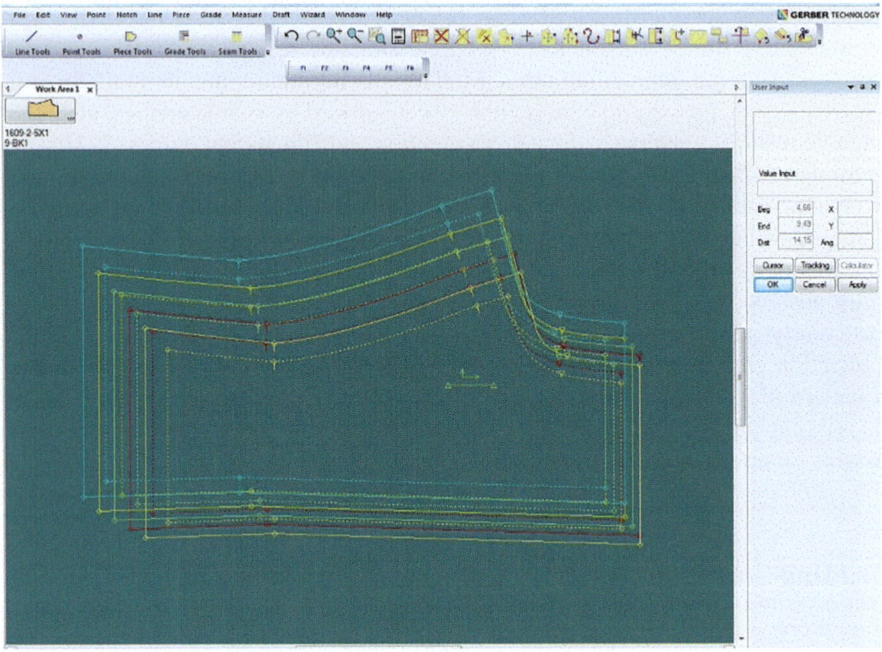

Fig. 10.11 Grading of a basic pattern to create different size

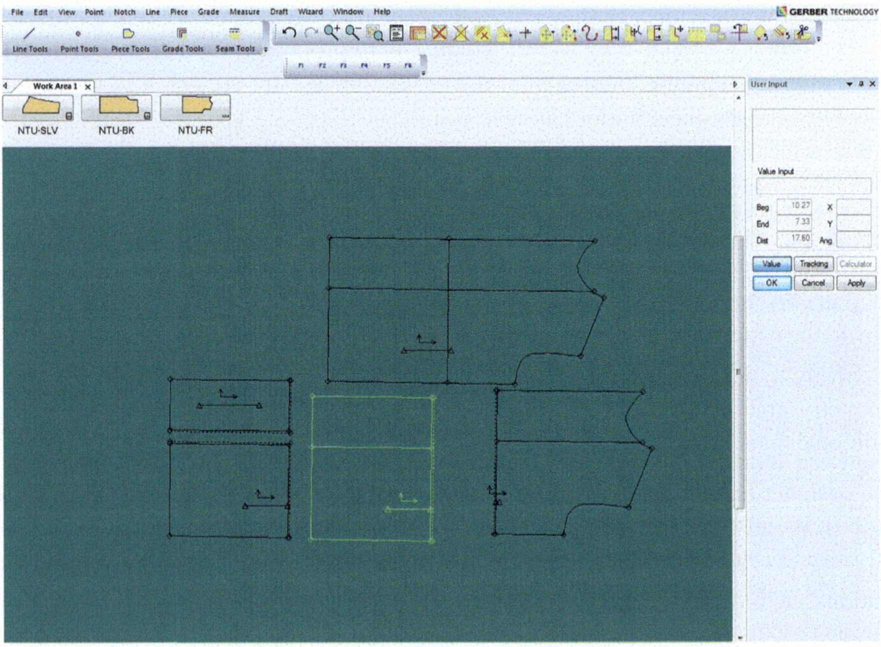

Fig. 10.12 Splitting of a basic pattern for grading according to design

PAD System

PAD system is known for its pattern grading capabilities. It helps in scaling patterns up or down to create multiple sizes. PAD system stands out as a premier garment CAD system in the clothing industry and fashion education, specializing in pattern-making and marker design. The cloud platform, Padshare, enhances collaboration by providing cooperative tools that leverage the advantages of cloud computing. Accelerate the patternmaking process with specialized tools, incorporate production details for mass production, and optimize fabric use through automated marker placement. This powerful yet user-friendly software is adept at pattern designing, grading, pattern alteration, and cloning, making it suitable for various production units. Developed by iWork Ltd, a Hong Kong-based company, PAD system offers a user-friendly interface that fosters innovation. It enables accurate computation and drawing of patterns by digitizing or photographing paper patterns. The PAD pattern design component facilitates swift design, drawing, and modification of patterns. For users requiring access to old patterns or comparing them with those imported from other CAD systems, the system allows retrieval of stored patterns from a frequently used database [23].

Streamlining the process of repeated pattern modifications for shapes shared across different fabrics such as self, lining, fusing, and facing, this system significantly enhances productivity. It simplifies the modification of the master pattern, allowing updates to be applied to its cloned pieces along with seam allowances. Individual cloned pieces can be independently locked and modified without impacting the master pattern. Users have the flexibility to make changes to both the master piece in the plan view window and the cloned pieces in the piece view window for various fabric types, ensuring consistency across elements like self, lining, and fusing. Efficiency is further increased as two or more pieces can be seamlessly combined to appear sewn together, eliminating the need for tedious individual piece changes. The synchronizing line feature enables the simulation of closed dart shirring, and merging pieces facilitates adjustments to the contour for a desired continuous curve. The tool simplifies the creation and modification of pleat, dart, and shirring characteristics with just a few clicks. Additionally, users can quickly and easily apply and adjust seam allowances, special corners, curves, and notches [23].

PAD automatic pattern grading has undergone continuous improvement, catering effectively to user needs. The software provides a range of grading tools, including symmetry grading, rotate grading, and the intuitive arrow grading, which mimics traditional manual grading but is easy to learn. With arrow grading, the manual input of X, Y coordinate values for point-to-point grading is replaced, making blending between points more spontaneous. By inputting grading values on the pattern's outer edge, PAD can automatically evaluate corresponding measurements for style lines inside the pattern. Specific measurements can be entered to establish clothing specifications and formulas for sewn segments, enabling the creation of a "frame window" that can be exported to Excel or a text document. This feature facilitates bulk production for tailor-made and custom-fit garments, aiding tailors and manufacturers in

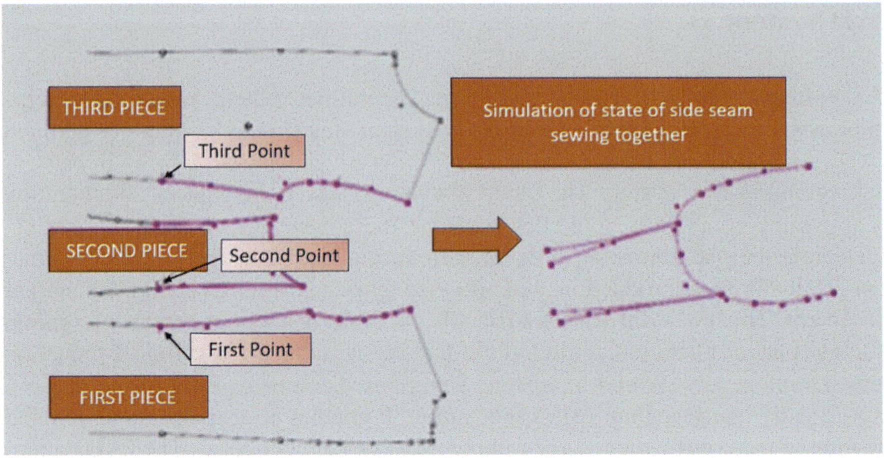

Fig. 10.13 Simulation of the panel within the pattern by using PAD system

efficiently crafting apparel to fit individuals or meet specific manufacturer specifications. The system utilizes grading values to automatically compute grading by inputting movement numbers from tape measure circumferences for each size. This approach establishes co-relations between patterns, ensuring the coherence of seams on connecting parts. This interconnectedness is key to PAD's remarkable speed in pattern processing, modifications, and amendments [23] (Fig. 10.13).

Tukatech

Tukatech offers a suite of software programs dedicated to marker creation and pattern grading, contributing significantly to maximizing fabric consumption during manufacturing. TUKA APM, a module within TUKAcad, stands out as the world's first fully automatic program for pattern creation and grading. Serving as a comprehensive CAD program for marking, grading, and pattern creation, it plays a foundational role in initiating the production of clothing products. TUKA APM is at the forefront of 3D clothing fit and virtual fashion design technology in the fashion industry. It boasts sophisticated features, including real-time motion simulation and personalized virtual fit models. As a web-based digital sample room, it acts as a miniature process life cycle management (PLM) system, bringing together the entire product development team under a unified virtual roof. This integrated approach enhances collaboration and coordination throughout the product development lifecycle [24] (Fig. 10.14).

Fashion technology provider, Tukatech, has introduced what it proclaims is the world's first fully automized patternmaking and grading software. This groundbreaking solution creates 2D CAD patterns based on measurements, including grade

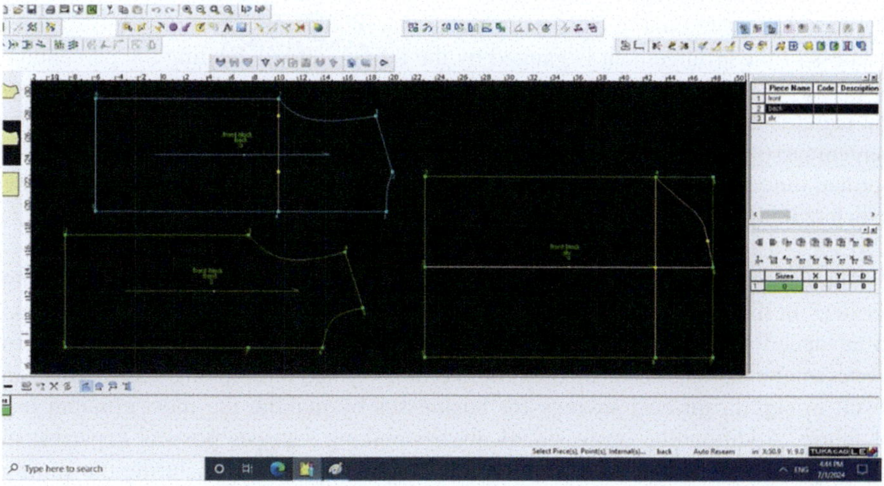

Fig. 10.14 Creation of digitized pattern by using Tukatech

rules, and provides costing markers for quotations. The newly exposed TUKA APM enables users to complete a graded pattern within a couple of seconds. In TUKAcad, digital patternmakers can choose any basic pattern from a library of silhouettes, including T-shirts, trousers, polos, dresses, and other apparel items for men, women, and children. For each selected silhouette, patternmakers feed data of different points of measurement, to get the illustration of the garment. The automated tool then allows patternmakers of any skill level to spontaneously create an initial 2D CAD pattern from these measurements and instantly apply grading for larger and smaller sizes [24].

Marker-Making Software

Marker-making software is employed to create layouts that are optimal for cutting fabric, reducing waste, and enhancing productivity in garment manufacturing processes. Once patterns are created through tools or pattern alteration from digitization, the subsequent step is grading. Following grading, the next critical process is marker making. Marker efficiency, which is the percentage of total fabric effectively utilized in garment parts, is automatically determined by marker-making software. The significance of marker efficiency lies in its direct connection to fabric cutting and the potential for fabric waste during production. As fabric utilization directly impacts costing, software companies continuously strive to incorporate special features that ensure maximum efficiency. Marker making plays a pivotal role in the apparel industry, involving the creation of stencils to guide fabric cutting for garment production. The marker planner, also known as the marker maker guru, is responsible for organizing patterns in a way that minimizes fabric waste. The effectiveness of the arrangement is gauged by the fabric yield, indicating the percentage of fabric cut into pieces. A high fabric yield implies a tight marker, signifying that the patterns are closely arranged. Fabric utilization, on the other hand, denotes the percentage of fabric genuinely utilized in the production of the garment. To ensure optimal outcomes, our marker-making experts leverage CAD systems, contributing to achieving high yields and tight markers, ultimately resulting in increased profits [25].

There are numerous advantages to employing professional marker-making services in the garment manufacturing process. Firstly, these services have access to advanced software and technology that can optimize the placement of pattern pieces on the marker, ensuring maximum efficiency and minimizing waste. This can result in significant cost savings for businesses by making the most efficient use of fabric. Additionally, professional marker-making services possess expertise in creating accurate and precise markers, reducing the risk of errors during the cutting process. This, in turn, leads to improved quality and consistency in finished products. Furthermore, outsourcing marker making to professionals allows businesses to focus on other aspects, saving time and resources. In summary, investing in professional marker-making services can enhance efficiency and accuracy in the garment manufacturing process, ultimately leading to improved profitability and customer satisfaction [25] (Fig. 10.15).

Lectra Diamino Fashion

When constructing markers, Diamino Fashion is utilized to assist create markers with the least amount of wasted cloth. Modaris, developed by Lectra, is patternmaking software that empowers fashion companies to create patterns and production-ready prototypes. It incorporates best practices in product development, ensuring a consistent level of quality and excellence throughout the entire product cycle. Among the versions available, Modaris Expert stands out as the most advanced in patternmaking technology. Modaris Expert significantly accelerates the product development process by allowing patternmakers to leverage their expertise. It streamlines the pattern creation process by error-proofing tasks and eliminating repetitive elements, leading to a potential saving of up to 30% in pattern adjustment time, thanks to its robust parametric capabilities. With Modaris Expert, users can achieve a higher level of efficiency compared to Modaris Classic [21].

Modaris 3D seamlessly synchronizes 2D production patterns with their 3D prototypes, enabling adjustments to be simultaneously applied to both 3D and 2D components, ensuring a perfect fit for garments of all sizes. Modaris 3D offers unique tools for verifying pattern fit, an extensive fabric library categorized based on characteristics, and other valuable features. In contrast, Modaris Classic serves as a versatile tool, allowing users to create, modify, grade, and industrialize patterns. It facilitates

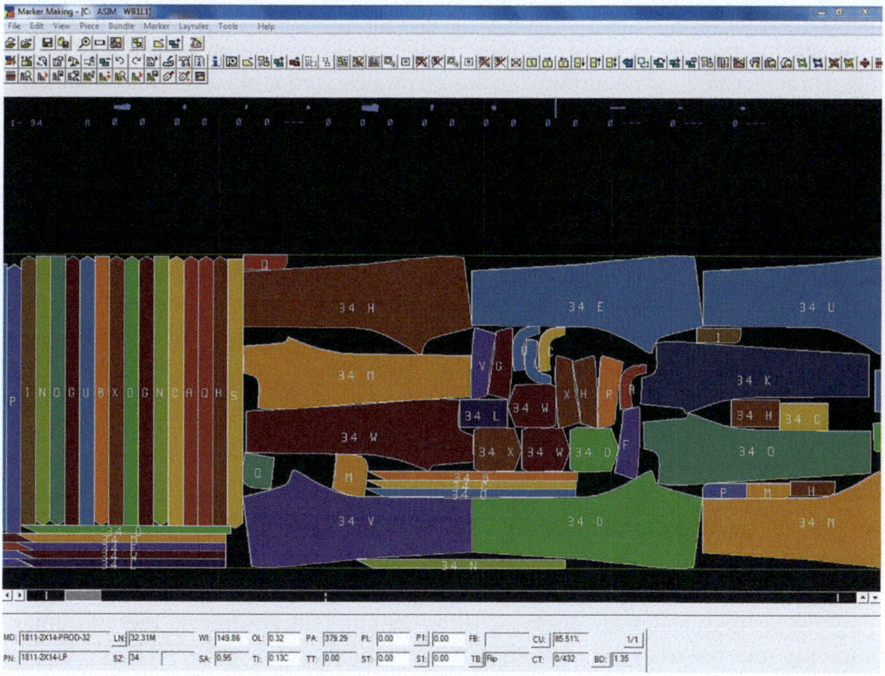

Fig. 10.15 Marker making of a digitized pattern for maximum efficiency

the development and storage of graded base patterns in a library, offering the convenience of using them for future models to expedite the overall process. For grading and industrializing patterns saved in native Modaris files, Modaris Essential proves to be an ideal choice. Particularly beneficial for suppliers during pre-production stages, this version enhances collaboration with brands by enabling the seamless sharing of pattern files without any risk of data loss or errors [21].

CLO 3D

Creating realistic 3D clothing, simulating fabric behavior, and conceptualizing ideas on virtual models are all made possible by the well-known 3D fashion design program CLO 3D [12]. CLO 3D, developed by CLO Virtual Fashion Inc., is a 3D tool specifically designed for designing and conceptualizing clothing and accessories. The primary market for this virtual 3D clothing prototype and simulation tool is the fashion and apparel industry. Within CLO 3D, designers have access to a number of tools and characteristics to create life-like 3D clothes. The study targets to discover the merits of CLO 3D and compare them with more conventional manual pattern-cutting approaches. This program employs complex algorithms and physics-based models to replicate the behavior of textiles and their connection with the body. With the increasing popularity of 3D printing in the world of fashion industry, CLO 3D plays a crucial role in streamlining the design process, reducing costs, and enhancing interaction and conceptualization of garment ideas. It has become an essential tool for fashion designers, patternmakers, and garment manufacturers, allowing them to bring their designs to life in a simulated environment [26].

Furthermore, there is an emphasis on maximizing the utilization of CLO 3D within the fashion and textile sectors to address ecological footprint and promote sustainability. CLO 3D not only enhances overall efficiency but also elevates competitiveness within the industry. Patternmakers, seeking to enhance their skills, performance, and contributions to the product development process, may choose to undergo training in CLO 3D. Businesses can also assess the return on investment (ROI) associated with integrating CLO 3D into their product development workflows. In the fashion industry, digital pattern production is gradually substituting traditional hand-drawn designs on drafting paper as the more contemporary and selected method. The ability for users to visualize the final design within the digital prototype while creating a digital pattern contributes to a precision level that is up to 30% more accurate and efficient in terms of both time and cost [26].

The process of creating a 3D model from a 2D digital pattern, utilizing 3D technology in fashion design by applying different colors according to the concept, is commonly referred to as 3D rendering. Among the most advanced technologies for rendering, CLO 3D proves to be specifically valuable. This technology empowers marketers to make instant design changes and assess the outcomes efficiently. By examining fit and silhouette before the actual design process, the overall design quality can be significantly enhanced, leading to reduced waste and errors occurred in the sampling phase before completion. CLO 3D serves as a valuable tool in transforming 2D drawings into 3D graphics within the virtual design program. It proves instrumental in evaluating fit, visualizing designs in practical use, and experimenting with various fabric and color options. Additionally, CLO 3D allows the export of patterns in various formats. The primary objective is to understand the current landscape of manual patternmaking and the fusion of 3D fashion design software in the apparel industry. The analysis of the CLO approach encompasses considerations related to input, output, control systems, human resources, technology, time, decision-making, and communication systems [26].

The simulation function in CLO 3D provides pattern creators and designers with the possibility to evaluate and assess their work within a digital environment, enhancing their understanding of the appearance and behavior of garments. This functionality contributes to improved fit and drape, optimizing processes, and reducing the costs associated with physical prototypes. Furthermore, it enhances accuracy, efficiency, uniqueness, and innovation in garment design. The process of converting a virtual model into a 2D or 3D image or animation of a garment design in CLO 3D refers to rendering. This feature allows pattern designers and producers to generate precise visuals of their work and complete with lighting, shading, and texture. The resulting visuals are valuable for marketing, communication with stakeholders, and design evaluation. Rendering plays a vital role for patternmakers, designers, and

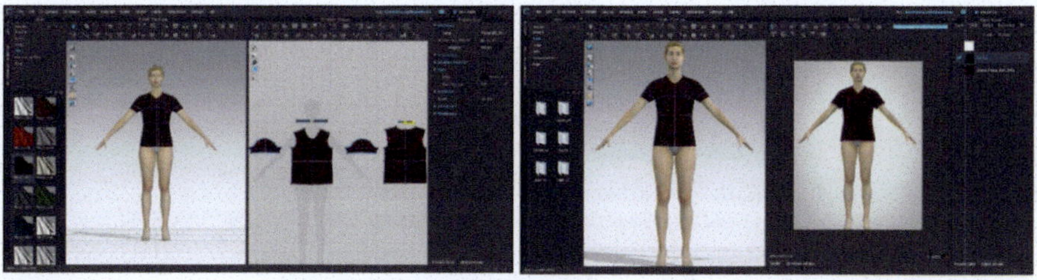

Fig. 10.16 Simulation and render window in CLO 3D

product manufacturers in the fashion industry as it provides an accurate and realistic perspective of the garment, enabling them to detect and address fit and drape issues effectively [26] (Fig. 10.16).

Browzwear

Designers may generate, visualize, and evaluate ideas in a virtual 3D environment with the help of Browzwear 3D solutions for fashion design and development [12]. With the continuous advancement of 3D tools, users can now go online and seamlessly share information with manufacturers regarding garments and their functionalities. The availability of textile textures, wraps, wefts, garment movements, and various textile behaviors in different conditions, such as water resistance, eliminates the need for multiple exchanges of prototypes [27]. Browzwear is revolutionizing clothing procedures by enabling the efficient creation of real-life 3D visualizations that can be shared and deployed across the entire value chain, ensuring comprehensive adaptability. Leveraging a true-to-life digital prototype allows for the generation of infinite variations of designs, which can be instantly shared throughout the organization. This facilitates efficient workflow execution and informed decision-making by conducting precise fit and style authentication using a digital twin of the physical garment [28].

Showcase apparel collections in captivating 3D catalogs and virtual showrooms, animating 3D salesman samples to capture intricate details and movements. Browzwear's 3D fashion design software facilitates the creation of comprehensive production-ready tech packs, considering both physical and visual properties of the garment. The virtual prototypes provided by the software accurately mirror real-life products, streamlining the sampling and production processes. This results in the swift introduction of ideas to the market, while simultaneously reducing costs and waste at every stage of the process. Fashion industry leaders and manufacturers worldwide rely on Browzwear's 3D fashion design software to expedite product development [28].

Utilizing these tools provides users with complete visibility and access to the patterns of a garment. Consequently, the reaccreditation and modification of garment

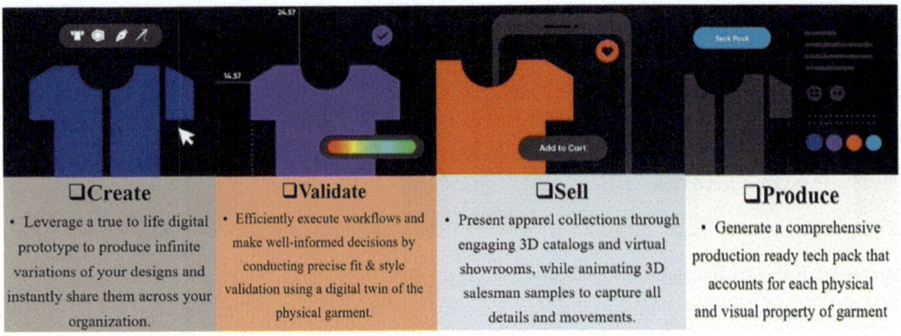

Fig. 10.17 Versatility of Browzwear in different ways

patterns can be carried out digitally, eliminating the need for physical prototyping. Established in 1999, Browzwear stands as a groundbreaker in providing 3D digital solutions for the fashion industry, facilitating flawless and sustainable processes from concept to trade. For designers, Browzwear expedites collection development, offering unlimited possibilities for creating various versions of styles. Technical designers and patternmakers benefit from Browzwear by quickly fitting graded garment patterns to any model with accurate, true motion, and material replication. Manufacturers find Browzwear's tech pack invaluable, as it ensures a smooth workflow from design to production by providing everything needed to produce physical garments perfectly the first time [29] (Fig. 10.17).

Conclusion

Patternmaking plays an essential role in shaping the conceptual design into clothes for numerous applications. In the patternmaking industry, both traditional and virtual pattern building come with their own set of advantages and drawbacks. The utilization of software for pattern manufacturing offers numerous benefits to the fashion industry, including accelerated iteration processes, enhanced design accuracy, replication and rendering, among other merits. However, there are certain pitfalls to consider when using these programs. Challenges may arise for small businesses or individual designers due to initial expenditures, learning curves, and dependence on technology, making it challenging to validate the investment. Another noteworthy limitation is the potential lack of feedback and limited understanding of conventional pattern construction, as designers may entirely rely on digital modeling for making design decisions. Furthermore, there is concern that the use of this software may diminish the creativity and originality of fashion designers, leading to designs that are too similar to each other. Despite these concerns, this software remains a valuable tool for patternmaking in the fashion industry. Designers and producers should carefully evaluate the pros and cons of the program to determine its suitability for their specific needs. However, one major disadvantage of conventional patternmaking is its laborious and time-intensive nature, with designs taking a considerable amount of time to develop. Therefore, traditional approaches may be lagging for manufacturing projects with short lead times. Moreover, traditional patternmaking heavily relies on hand-operated processes, which can lead to more errors and inconsistencies. The precision and accuracy provided by 3D software surpasses what traditional patternmaking can achieve. Even skilled patternmakers may encounter factors and elements that impact achievements. Conclusively, traditional, and digital approaches can be employed to produce elegant, advanced, and sophisticated designs that meet the expectations of modern consumers and have a crucial role in the apparel industry.

References

1. S. Efrat, *The Development of a Method for Generating Patterns for Garments that Conform to the Shape of the Human Body* (1982)
2. D.B. Datta, P. Seal, Various approaches in pattern making for garment sector. *J. Text. Eng. Fashion Technol.* **4**(1) (2018). https://doi.org/10.15406/JTEFT.2018.04.00118
3. S. Gill, N. Chadwick, Determination of ease allowances included in pattern construction methods. Int. J. Fashion Des. Technol. Educ. **2**(1), 23–31 (2009). https://doi.org/10.1080/17543260903018990
4. L. Perkins, Transparency considerations for the ethical fashion designer, in *Original Citation Perkins* (2021)
5. A.S.M. Sayem, R. Kennon, N. Clarke, 3D CAD systems for the clothing industry. Int. J. Fashion Des. Technol. Educ. **3**(2), 45–53 (2010). https://doi.org/10.1080/17543261003689888
6. P. Jin et al., Design and research of automatic garment-pattern-generation system based on parameterized design. Sustainability (Switzerland) **15**(2), 1268 (2023). https://doi.org/10.3390/SU15021268/S1
7. A.S.M. Sayem, *Digital Fashion Innovations: Advances in Design, Simulation, and Industry* (Jan. 2023), pp. 1–206. https://doi.org/10.1201/9781003264958
8. Y. Jhanji, *Computer-Aided Design—Garment Designing and Patternmaking* (Technological Institute of Textile & Sciences, Bhiwani)
9. Pattern Making in Fashion Design—Arch College of Design & Business Blog. [Online]. Available: https://www.archedu.org/blog/pattern-making-in-fashion-design/. Accessed 17 Jun 2024
10. 3 Methods of Garment Pattern Making Online—Hamstech Online Blog. [Online]. Available: https://www.hunarcourses.com/blog/3-methods-of-pattern-making/. Accessed 17 Jun 2024
11. Essentials of Pattern Making and Why It's Important to Get it Right! | Sewport. [Online]. Available: https://sewport.com/learn/pattern-making. Accessed 17 Jun 2024
12. Software Used for Pattern Making—Textile Learner. [Online]. Available: https://textilelearner.net/software-used-for-pattern-making/. Accessed 17 Jun 2024
13. Textile Design Software | CAD & CAM—Lesson | Study.com. [Online]. Available: https://study.com/academy/lesson/how-is-cad-cam-used-in-textiles.html. Accessed 17 Jun 2024
14. How the Fashion Industry Uses Illustrator | Classes Near Me Blog. [Online]. Available: https://www.nobledesktop.com/classes-near-me/blog/illustrator-in-fashion. Accessed 17 Jun 2024
15. Enhancing Fashion Design with Adobe Illustrator & Photoshop—Chitkara University. [Online]. Available: https://www.chitkara.edu.in/blogs/enhancing-fashion-design-with-adobe-illustrator-photoshop/. Accessed 17 Jun 2024
16. Benefits of CorelDraw for Business | CorelDRAW Graphics Suite 2020. [Online]. Available: https://www.wroffy.com/benefits-of-coreldraw-for-business/. Accessed 17 Jun 2024

17. Y. Wang, Digitalization of garment design based on corelDRAW software. *Comput. Aided Des. Appl.* **17**(2), 111–122 (2020). https://doi.org/10.14733/cadaps.2020.S2.111-122
18. Setting Clothes in Motion with Israel's OptiTex—ISRAEL21c. [Online]. Available: https://www.israel21c.org/setting-clothes-in-motion-with-israels-optitex/. Accessed 17 Jun 2024
19. Gerber Garment Technology (GGT). [Online]. Available: https://www.ambition-apparel.com/pattern-making.php. Accessed 17 Jun 2024
20. Gerber AccuMark CAD Software for Fashion Design | Lectra. [Online]. Available: https://www.lectra.com/en/products/gerber-accumark-accunest-fashion. Accessed 17 Jun 2024
21. Modaris: Smart, Speedy and Efficient Patternmaking for Fashion Experts | Lectra. [Online]. Available: https://www.lectra.com/en/products/modaris-expert. Accessed 17 Jun 2024
22. What is Pattern Grading in the Fashion/Garment Industry. [Online]. Available: https://techpacker.com/blog/design/pattern-grading-in-the-fashion-garment-industry/. Accessed 17 Jun 2024
23. Padsystem—Pattern-Aided Design (PAD) System. [Online]. Available: https://www.padsystem.com/en/. Accessed 06 Feb 2024
24. Tukatech Unveils Fully Automatic Pattern Making System—Just Style. [Online]. Available: https://www.just-style.com/news/tukatech-unveils-fully-automatic-pattern-making-system/. Accessed 17 Jun 2024
25. Marker Making: How Professional Marker Making Services Maximizes Effic. [Online]. Available: https://www.smartpatternmaking.com/pages/marker-making-how-professional-marker-making-services-maximizes-efficiency. Accessed 17 Jun 2024
26. Md.A. Habib, Md.S. Alam, A comparative study of 3D virtual pattern and traditional pattern making. J. Text. Sci. Technol. **10**(01), 1–24 (2024). https://doi.org/10.4236/JTST.2024.101001
27. E. Papahristou, N. Bilalis, 3D virtual prototyping traces new avenues for fashion design and product development: a qualitative study. *J. Text. Sci. Eng.* **07**(02) 2017. https://doi.org/10.4172/2165-8064.1000297
28. Browzwear: #1 3D Apparel Design Software. [Online]. Available: https://browzwear.com/. Accessed 17 Jun 2024
29. Superdry X Browzwear. [Online]. Available: https://browzwear.com/success-stories/superdry. Accessed 17 Jun 2024

Chapter 11
Introduction of Marker and Types

Saeed Akhtar, Muhammad Qamar Khan, and Madeha Jabbar

Keywords Marker making · Grain lines · Marker efficiency · Fabric wastage · Fabric utilization · CAD

What is a Marker?

Marker making is essential in garment production to minimize fabric waste and optimize efficiency. A marker serves as a blueprint guiding the cutting of fabric into garment pieces. It comprises a precise arrangement of pattern pieces for various sizes of a specific style to be cut from a single spread.

Essentially, a marker is a strategic arrangement of pattern pieces on fabric that minimizes waste while ensuring desired grainlines are maintained. It functions as a template from which all pieces are cut, laid atop fabric layers, and then cut accordingly [1] (Fig. 11.1).

What is Marker Making?

Marker making involves determining the most efficient layout of pattern pieces for a particular style, fabric, and size distribution.

In factories without CAD systems, this process is carried out manually, requiring time, skill, and concentration. However, modern marker making is predominantly

S. Akhtar (✉) · M. Q. Khan · M. Jabbar
Department of Textile Engineering, School of Engineering and Technology, National Textile University Faisalabad, Faisalabad, Pakistan
e-mail: saeed.akhtar@ntu.edu.pk

M. Q. Khan et al. (eds.), *Garment Sizing and Pattern Making*, SDGs and Textiles,
https://doi.org/10.1007/978-981-97-7683-2_11

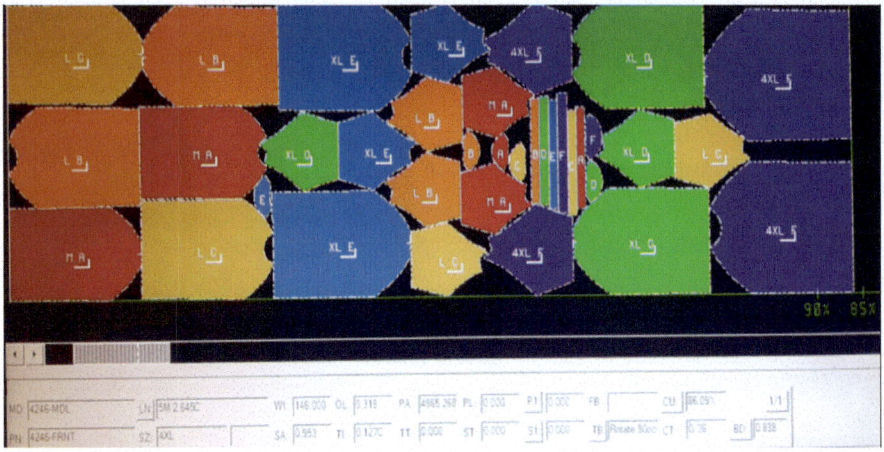

Fig. 11.1 Marker

facilitated by CAD systems, offering precision, enhanced control, and reduced time requirements.

Typically, a marker refers to a paper document containing all size patterns of a garment, enabling the production of the required number of garments with minimal fabric consumption. Marker width corresponds to the minimum fabric width, while its length varies based on the number of pattern sizes used and other considerations such as [2]:

- Number of garments per fabric lay
- Length of the cutting table
- Production planning

Before marker making, several factors need to be considered:

- Fabric width must exceed marker width by at least 1/2 inch.
- Fabric length must exceed marker length.
- Length of the cutting table.
- Production planning.
- Alignment of the grain line parallel to the warp line in woven fabric or wale in knitted fabric when laying the marker on the fabric piece.

Maker Making Flowchart

See Fig. 11.2.

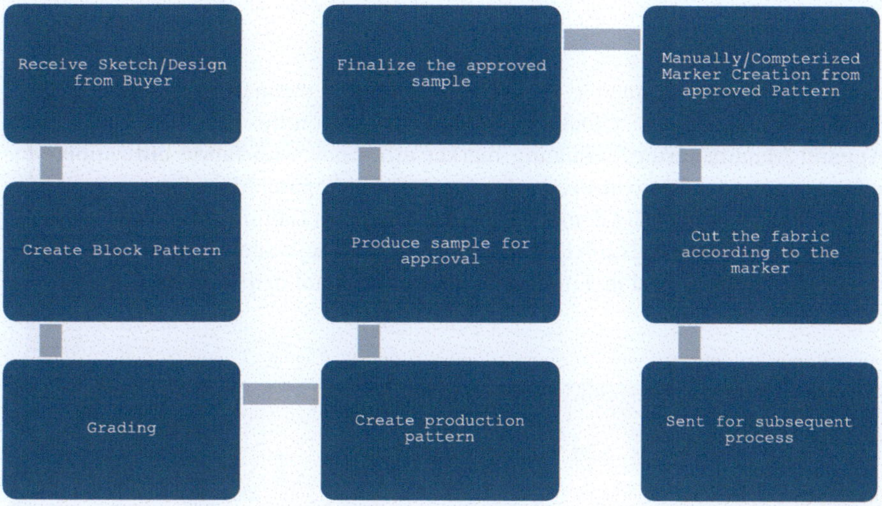

Fig. 11.2 Flowchart for marker making

Importance of Marker Making

Marker making holds significant importance in the garment manufacturing process. It serves to minimize fabric consumption and wastage, consequently enhancing production efficiency. Efficient marker making facilitates grading, ensuring various garment components adhere to standard body measurements.

This not only saves time but also cuts down costs. Within the fashion and garment manufacturing industry, meticulous creation of patterns and templates for garment components before production is paramount. This step plays a pivotal role in guaranteeing the quality, fit, and cost-effectiveness of the final product [2].

Objectives of Marker Making

1. Minimize fabric wastage during cutting.
2. Expedite bulk fabric cutting.
3. Enhance cutting efficiency.
4. Facilitate grading processes.
5. Optimize pattern-making benefits.
6. Maintain cutting precision.
7. Reduce wastage and time, thereby boosting overall efficiency and production.

Marker Efficiency

Marker efficiency is defined as the ratio of the area of the marker used in a garment to the total marker area. Glock and Kunz elaborate on this concept in their book "Apparel Manufacturing," equating marker efficiency with fabric utilization—the percentage of total fabric utilized in garment parts (Apparel Manufacturing: Sewing Product Analysis, Ruth E Glock and Grace I Kunz, 4th Edition 2004, Prentice Hall). The success of a marker's job is gauged by its efficiency, calculated as the occupied area by the patterns used in a marker relative to the total marker area [2] (Figs. 11.3 and 11.4).

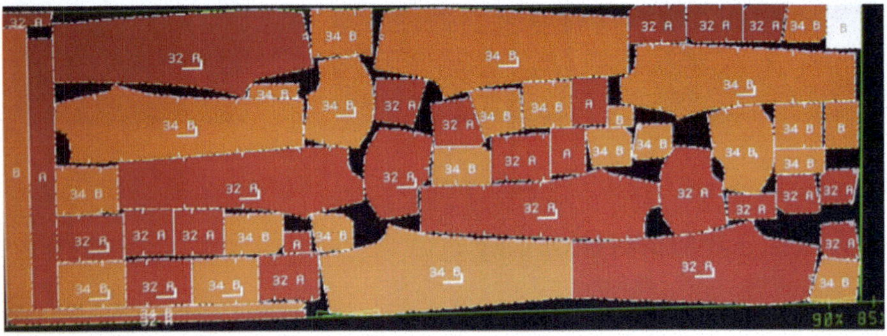

Fig. 11.3 Marker efficiency for Denim at 86%

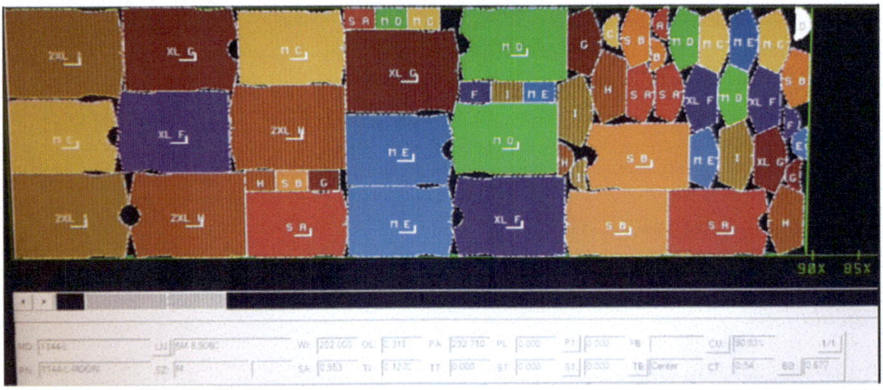

Fig. 11.4 Marker efficiency for knitted garment (Polo shirt) at 91%

Determining Marker Efficiency Objectives

Several key objectives underpin the determination of marker efficiency:

1. Optimal fabric utilization.
2. Minimization of fabric wastage.
3. Maximization of profits from garment export orders.
4. Identification of factors influencing material utilization.

Formula 1

Marker Efficiency = (Area of patterns used in the marker/Area of the marker) × 100

In CAD markers, the system automatically calculates the area of garment pattern pieces placed within the marker, while the total marker area is derived by multiplying marker length by marker width. For manual markers, measuring the surface area of garment patterns can be challenging. Mechanical devices like a planimeter may be employed for this purpose.

Formula 2

Marker Efficiency% = (Weight of fabric consumed by pattern pieces in a marker/ Total weight of fabric under the marker area) × 100

To compute marker efficiency using this formula, weigh garment parts cut from one layer of fabric according to markers and determine the weight of fabric (one layer) covering the total marker area. Higher marker efficiency correlates with lower fabric wastage, ultimately reducing the amount of fabric required for a specific number of garments. Given that approximately 50% of a garment's cost is attributed to fabric, even a 1% increase in marker efficiency can significantly impact profits. Thus, marker making stands as a crucial task within the apparel manufacturing process.

Note: *Marker efficiency and fabric utilization are distinct; the former does not account for fabric wastage due to end bits and end loss, unlike fabric utilization calculations.*

Factors Which are Involved with the Marker Efficiency

Factors influencing marker efficiency encompass various elements including the marker planner, garment size, marker length, marker width, pattern engineering, fabric characteristics, and marker-making method [3].

(a) **Marker Maker**

Marker efficiency is predominantly contingent upon the expertise, diligence, integrity, experience, and technical acumen of the marker personnel. Greater repetition in marker creation typically leads to improved efficiency.

(b) **Garment Size**

In marker making, incorporating a broader range of garment sizes generally enhances marker efficiency. However, excessively diverse size variations may diminish efficiency.

(c) **Marker Length**

Marker length is influenced by multiple factors. Longer markers typically correlate with higher efficiency and increased cutting room productivity. Conversely, excessive marker length may yield adverse outcomes.

(d) **Pattern Engineering**

Marker efficiency can be bolstered through alterations to pattern designs, such as pattern division or fragmentation.

(e) **Fabric Characteristics**

Fabric symmetry significantly impacts marker efficiency. Symmetrical fabrics tend to yield higher efficiency, while asymmetrical fabrics may result in reduced efficiency, particularly when aligning stripes or checks.

(f) **Marker Making Method**

Marker creation methods, including manual and computer-aided approaches, affect efficiency. Computer-aided markers generally offer superior efficiency, albeit at the expense of time. However, markers produced by experienced personnel can achieve high efficiency levels in less time.

(g) **Marker Width**

Marker width is typically determined by fabric width. Studies indicate that wider markers facilitate easier planning for marker personnel, thereby increasing efficiency. However, this principle may not apply to tubular fabrics.

(h) **Garment Type**

Marker efficiency varies depending on garment type. Garments with larger size patterns relative to smaller sizes may exhibit lower efficiency due to increased fabric wastage. Conversely, garments with a higher proportion of smaller size patterns typically achieve higher efficiency during marker creation.

In essence, marker efficiency is influenced by a multitude of factors, each playing a crucial role in optimizing fabric utilization and minimizing wastage during the marker-making process.

Constraints of Marker Making

To achieve an effective marker, a system facilitating clear visualization of the marker's complete dimensions is essential for the marker personnel. Typically, larger patterns are initially positioned, followed by smaller ones fitted into the spaces between them. This approach minimizes fabric wastage, thereby enhancing marker efficiency. Additionally, the marker personnel may strategically rotate or reposition patterns to reduce the marker's length, further optimizing efficiency.

(a) **Grain Line**

Each pattern is delineated with a grain line indicating the warp side of the fabric. During marker making, it is imperative to align the pattern's grain line parallel to the fabric's warp side in woven fabrics and wales in knitted fabrics. Depending on garment designs, patterns may occasionally be positioned at 45-degree angles. The drape and hang of garments hinge on the alignment of the fabric's grain line with the pattern's grain line. To enhance marker efficiency, the fabric's grain line can be positioned at a 2-degree angle with the pattern's grain line. However, caution must be exercised to ensure that garment designs are not significantly affected. Monitoring grain line alignment serves as a constraint for the marker maker.

(b) **Fabric Characteristics**

Certain fabrics exhibit symmetrical properties when rotated at a 180-degree angle, known as symmetrical fabrics. Marker making poses no challenges with symmetrical fabrics. However, asymmetrical fabrics, such as pile fabric or certain printed fabrics, appear different when rotated at 180 degrees. Special attention is necessary during marker making for asymmetrical fabrics to prevent garment defects. Random movement of patterns is restricted during fabric spreading for asymmetrical fabrics, imposing a constraint on the process.

(c) **Garment Design Characteristics**

In garments requiring specific design attributes, such as mirror images, careful attention is crucial during fabric spreading to ensure consistent designs in each garment.

(d) **Cutting Quality**

Patterns are typically cut separately from the fabric lay using a knife blade, which may pose challenges in rotating the blade at a 90-degree angle. Precautionary measures are necessary during marker making to ensure smooth pattern cutting and avoid defective cuts. After marker creation, verifying the total number of patterns is essential to avoid fabric wastage or production delays. Assigning code numbers to each pattern facilitates easy separation after fabric cutting, minimizing errors and optimizing resource utilization.

Immediately following the completion of the marker, it is imperative to tally the total number of patterns. For instance, if producing pants in six sizes with sixteen patterns per size, the total count would be $6 \times 16 = 96$ pattern pieces. Any shortfall in the number of patterns during sewing after fabric cutting could result in wastage of time, money, and resources. Hence, it is crucial to conduct a thorough count of patterns right after marker creation. Assigning unique code numbers to each pattern facilitates easy identification and separation post-fabric cutting, preventing mixing of patterns from different sizes. Failure to implement this precaution could lead to compounding errors, ultimately resulting in significant wastage of time and resources.

Production planning involves organizing the execution of garment orders, typically provided in dozens, with specifications on size, color, and quantity ratios. For instance, consider an order for 1500 dozen ladies dresses, comprising 600 dozen white, 600 dozen blue, and 300 dozen cream garments, with size ratios of 2:4:4:2 for sizes 15, 16, 17, and 18, respectively.

Various strategies can be employed for production planning and marker creation. Initially, one approach involves crafting markers based on size ratios, such as 2 sets for size 15, 4 sets for size 16, 4 sets for size 17, and 2 sets for size 18, totaling 12 sets of patterns. While this method may yield higher marker efficiency, it demands considerable time for marker creation. Moreover, longer markers may pose challenges in accommodating fabric spreads on cutting tables and may require additional fabric rolls, potentially leading to increased handling costs and variations in garment shades due to fabric ply discrepancies.

Alternatively, the order can be split into two markers. The first marker could comprise 2 sets each for sizes 15, 16, and 17, while the second marker could include 2 sets each for sizes 16, 17, and 18, totaling six sets of patterns for each marker. This approach allows for more manageable marker lengths, facilitating swift production and suitability for smaller cutting tables. However, it may result in reduced marker efficiency.

The choice between these approaches underscores the intricate relationship between marker making and production planning.

Methods of Marker Making

After the patterns for garments in all required sizes are prepared, the marker-making process commences. Initially, the marker width is determined based on the minimum width of the fabric intended for garment production. Typically, markers are created on white or newsprint paper (Apparel Manufacturing: Sewing Product Analysis, Ruth E Glock and Grace I Kunz, 4th Edition 2004, Prentice Hall). Large patterns are placed first, followed by smaller patterns positioned in the spaces between the larger ones to enhance marker efficiency. Marker making is a highly skilled and technical endeavor, primarily executed through two main methods:

1. **Manual Marker Making**

This method involves physical and mental exertion. Manual markers can be created using either full-size patterns or reduced-size patterns, both of which are elaborated below (Fig. 11.5):

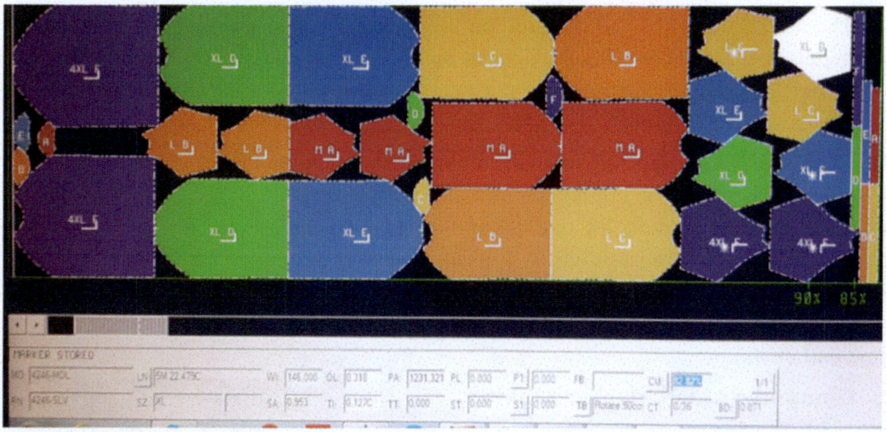

Fig. 11.5 Manual marker

(a) **Full-Size Patterns**

In this approach, patterns are in their original, full sizes according to standard measurements. The relatively rigid patterns are placed on paper or fabric, and the marker maker endeavors to arrange them optimally by maneuvering them in various directions to maximize marker efficiency.

This method is typically suitable for creating markers of shorter lengths, as it is easier for the marker maker to manage and control all patterns within a confined space. However, for long markers comprising patterns of various sizes, this method becomes challenging, as efficiently placing patterns along the entire length is arduous.

Marker-making table, upon which the paper or fabric is placed for marker creation, may have a fixed or tilting surface. A tilting table can be adjusted to various angles to facilitate marker making, and a vacuum system beneath the table surface helps secure patterns in place.

(b) **Reduced-Size Patterns**

In this method, markers are also crafted through physical labor. Initially, the full-length patterns are reduced to 1/5th scale or 1/25th scale using a pantograph. Subsequently, in the 1/5th scale, marker planning is executed on a smaller table. The process unfolds systematically as follows:

1. The full-length patterns are downscaled using a pantograph, typically onto plastic sheets or rigid board paper.
2. Marker planning with the reduced-size patterns is conducted on a small table measuring approximately 50 cm wide and 2 m long, marked with 1/5th scale on both sides for ease of manipulation.
3. After completing the marker, a photograph is taken. If time permits, the marker may be recreated, and another photograph is captured.

4. The area covered by the patterns on the marker is measured using a planimeter to determine marker efficiency based on the ratio of marker area to patterns area.
5. The photograph of the efficient marker, along with concise patterns, is carefully preserved in a cabinet.

Observing the photograph of an efficient marker made with reduced-size patterns, full-size markers can be created by placing the patterns on paper or fabric accordingly. Occasionally, reduced-size patterns may have faults due to inaccuracies in the pantograph, and in such cases, full-size markers produced based on the photograph may exhibit greater efficiency. While this method is suitable for bulk garment production, it may not be practical for smaller-scale production due to the associated time and cost.

2. **Computer-Aided Marker Making**

To initiate computer-aided marker making, the production patterns must first be stored in the computer's memory. Typically, digitizing or scanning methods are employed for entering the production patterns into the computer. While the full-size production patterns are used for digitizing or scanning, the computer stores them in reduced sizes for visualization on the screen.

Once all production patterns required for a garment are inputted into the computer memory, the necessary patterns for all sizes are generated using the pre-stored grade rule. The computer-aided marker maker typically utilizes a computer screen, keyboard, and light pencil. Instructions are then provided to the computer by the marker maker, specifying the marker width, pattern sizes to be used, and any obstacles to marker making, which the computer consistently adheres to during the process. In this approach, the computer can either autonomously create the marker or assist the marker maker in the process.

When the computer independently generates the marker, it is referred to as automatic marker making. In this scenario, the computer generates the marker itself based on preplanned instructions, resulting in the most efficient marker possible. However, this method requires more time as the computer utilizes permutation and combination methods to create the marker.

Alternatively, when the marker maker creates the marker with the computer's assistance, it is known as interactive marker making. In this method, all pattern pieces are displayed in reduced dimensions on the top part of the computer screen, with horizontal parallel lines indicating the marker's width. A vertical line marks the starting point of the marker, with the right side initially left open.

Data such as marker identification, length, and dynamically changing efficiency are displayed below the horizontal lines. To create the marker interactively, a data pen and tablet are utilized. The tablet serves as the display screen, with the data pen featuring a pressure-sensitive pointer. Patterns can be placed anywhere on the screen using the pen, and patterns to be removed are marked with the cursor. By lightly pressing the pen on the tablet and moving it, the pattern appears on the screen and can be positioned within the marker. Releasing the pen then fixes the pattern in place.

This process allows for the rapid creation of an efficient marker. Upon completion, marker details such as efficiency, length, width, and pattern count are instantly

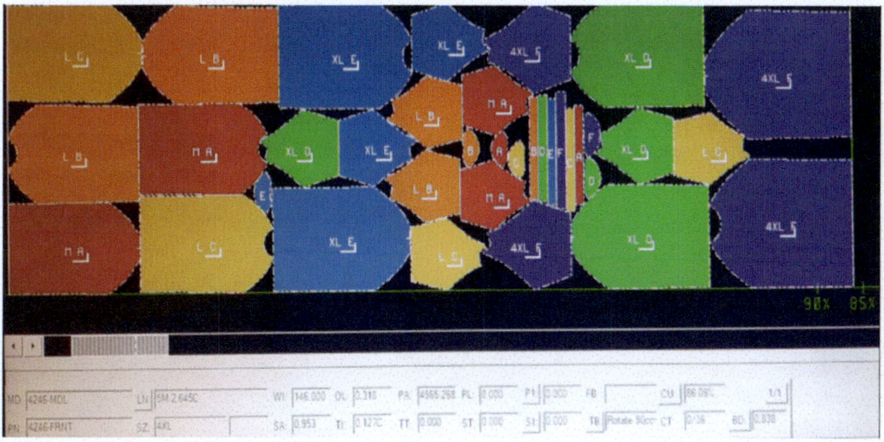

Fig. 11.6 Auto marker on CAD software

displayed on the screen. The completed marker can be saved in the computer memory and, if printing capabilities are available, a hard copy can be obtained.

Computer-aided marker making offers superior standard and efficiency compared to manual methods. It enables the fastest generation of markers and allows for the production of numerous copies as desired.

Overall, computer-aided marker making presents numerous advantages, although it requires a substantial initial investment. It can result in significant fabric savings, typically around 2% (Fig. 11.6).

3. **Mixed Marker**

A mixed marker remains the most favored marker type. In cases of garment asymmetry or the pursuit of optimal efficiency, the mixed marker is employed on fabric laid open and facing upwards on the table. For most woven fabrics and flat open knits, combined markers offer superior fabric utilization by placing patterns wherever they best fit along the marker's length.

4. **Open Marker**

Following closely, the open marker stands as the second most prevalent type. When dealing with asymmetric garments or aiming for enhanced quality during spreading, the open marker ensures that pairs of parts (left and right) are kept closely aligned along the marker's length. This marker type is also used on fabric spread open and facing upwards on the table, offering the second-best fabric utilization for most woven fabrics and flat open knits.

5. **Closed Marker**

The closed marker is reserved for specific situations. As only one-half of the pattern set is utilized, the fabric spread on the table is folded lengthwise and positioned face to face. Cutting through the fold results in obtaining both the left and right pieces of

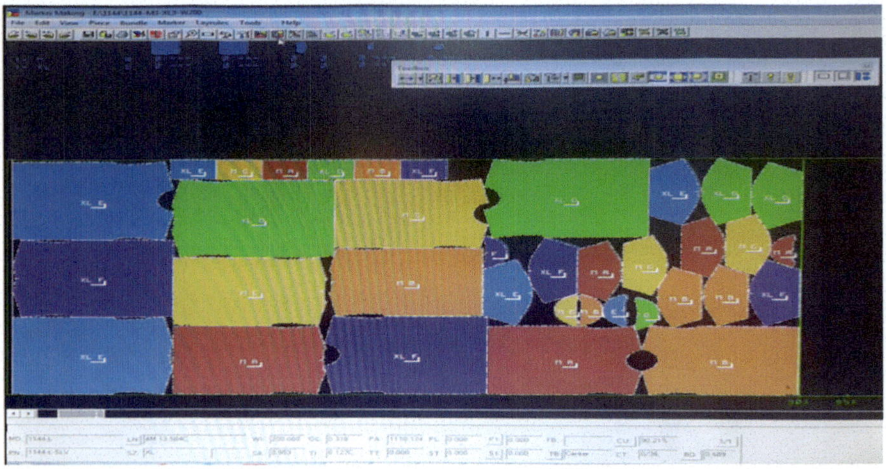

Fig. 11.7 Single section size marker

the garment from a pair of consecutive plies. Styles featuring pattern pieces covering the entire width, such as a one-piece back versus separate left and right back pattern pieces, can be oriented on the control fold of the marker. Cutting at the fold produces a one-piece part.

6. **Single Section Marker**

When patterns of various sizes are scattered throughout the entire length of the marker, positioned wherever they provide the best fit, the marker is labeled as non-sectional or single-section. This marker type boasts the highest fabric utilization (efficiency), as it maximizes the integration of pattern pieces, making it particularly advantageous for frequently ordered sizes (Fig. 11.7).

7. **Section Marker**

Markers with multiple sections, termed section markers, come into play when order ratios are unpredictable. Section markers facilitate the production of different order ratios using the same marker through step spreading. While offering lower efficiency, single-section markers provide total flexibility in accommodating diverse order ratios (Fig. 11.8).

Fabric Wastage Outside Marker

Fabric wastage outside the marker is primarily influenced by marker efficiency. However, additional fabric wastage can occur, sometimes surpassing marker planning control. To address such wastage, the following discussion outlines the factors contributing to this type of fabric wastage:

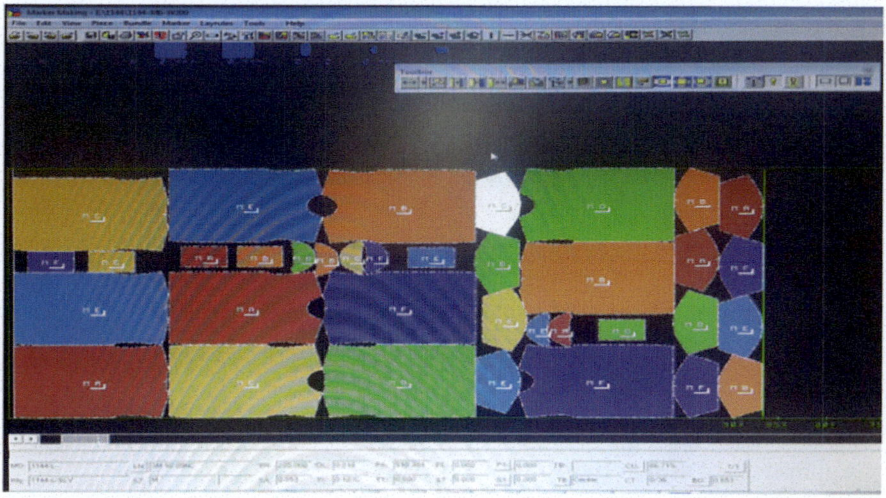

Fig. 11.8 Sectional marker

(a) **Ply End Losses**

Fabric extensibility and the constraints of fabric spreading machines necessitate allowing for an allowance at the end of each fabric lay. Typically, a 2 cm allowance is maintained on each side of the lay, resulting in a total of 4 cm of extended fabric per lay, which adds to the fabric wastage tally. The quantity of this type of wastage may vary based on fabric dimension stability. Thus, minimizing such wastage requires careful consideration of the appropriate allowance for each fabric.

Furthermore, larger marker lengths tend to decrease this type of wastage, while smaller marker lengths increase it.

(b) **Loss of Fabric Ends**

Fabrics in the garment industry typically come in rolls with specific length limitations. Fabric lays are based on the marker length, but discrepancies often arise in fabric lengths compared to lay lengths. Consequently, the ends of fabric rolls frequently result in splices or remnants, contributing to fabric wastage beyond marker planning. While this type of wastage cannot be entirely eliminated, it can be mitigated by using longer fabric rolls and strategically managing splice placement in the marker.

(c) **Selvedge Losses**

Each fabric contains selvedges along its width, often left out during cutting. Typically, this omitted quantity is considered as 3% of the fabric width, though for highly stable fabrics, it may be reduced to 2%. The extent of fabric width utilization depends on selvedge quantity, fabric width uniformity, and selvedge alignment during spreading. While widthwise fabric wastage falls outside marker planning, it can be minimized by utilizing wider fabrics.

(d) **Purchase Loss**

Fabrics are usually purchased based on length, with the length marked on the fabric roll by the manufacturer and supplier. However, discrepancies may arise where the actual fabric length falls short of the marked length. If undetected upon purchase, the buyer may incur losses due to fabric shortage. To mitigate this, fabrics should be thoroughly inspected before purchase and upon receipt of the supply.

Applications of Marker Making in Various Industries

Below, we delve into the diverse applications of marker making across different sectors [4]:

i. **Apparel and Fashion Industry**

Marker making serves as a cornerstone in the fashion and garment manufacturing domain, facilitating:

- **Clothing Production**: In apparel manufacturing, marker making arranges pattern pieces on fabric to minimize wastage and maximize garment output from fabric rolls, resulting in cost-effectiveness and eco-friendly production.
- **Variety of Garments**: Marker making contributes to the creation of patterns for a wide array of clothing items, including shirts, trousers, dresses, and outerwear.
- **Size Range**: It enables efficient production across various sizes and styles within clothing collections
- **Automotive Industry:**

Marker making holds significant importance in automotive interior manufacturing, particularly for upholstery and seat covers, ensuring:

- **Precision Cutting**: Marker making guarantees accurate cutting of fabrics, leather, and synthetic materials used in car seats, headliners, and interior panels.
- **Consistency**: It maintains uniformity in the appearance and quality of automotive interiors.
- **Home Textiles**

Marker making is indispensable for the production of diverse home textile products, such as:

- **Bedding**: Including bedspreads, duvet covers, and pillowcases.
- **Table Linens**: Such as tablecloths and napkins.
- **Bath Linens**: Encompassing towels and shower curtains.
- **Industrial Textiles**

The industrial sector relies on marker making for manufacturing various items, including:

- **Bags and Luggage**: Marker making ensures efficient material utilization for crafting bags, backpacks, and suitcases.
- **Tents and Awnings**: Marker making aids in the production of tents, canopies, and awnings for outdoor events and industrial purposes.
- **Protective Covers**: For equipment and machinery, marker making is employed to create covers that fit precisely.

Advantages of Marker Making

- Marker making contributes to minimizing fabric wastage by strategically arranging pattern pieces on the fabric layout. This is particularly crucial for mass production scenarios, where even slight reductions in fabric wastage can result in substantial cost savings.
- Marker making enhances production efficiency by streamlining the process of cutting garment pieces. With pattern pieces already optimally arranged, markers enable the cutting of multiple fabric layers simultaneously, saving time and effort.
- Marker making ensures the precision of the cutting process by consistently positioning pattern pieces accurately on the fabric. This is vital for maintaining high-quality standards and achieving consistent sizing in garments.
- Marker making offers versatility across various garment types, with grading providing advantages such as ensuring comfortable fits for individuals of all sizes. Additionally, grading aids businesses in enhancing production efficiency by reducing fabric waste and cutting time.

Disadvantages of Marker Making

- Computerized marker-making systems entail significant costs for both acquisition and maintenance.
- Skilled operators are required to effectively utilize computerized marker-making systems.
- Marker making can be time-consuming, especially when dealing with complex garment styles.
- Certain limitations exist for marker-making systems, such as challenges in accommodating certain fabric types or garment styles. These include napped fabrics, piled fabrics, stretch fabrics like jersey and spandex, fabrics with large patterns or designs, and fabrics with defects or imperfections.

Conclusion

As a textile engineer, I have closely observed the marker-making process. Based on my experience, I recognize marker making as a crucial process aimed at cost reduction, efficiency enhancement, accuracy improvement, and consistency maintenance. It is widely applicable across industries and involves rigorous quality control measures to ensure the production of high-quality markers.

Practically every garment and clothing manufacturing entity utilizes marker making to reap the associated benefits. Without marker making, these industries would likely encounter fabric wastage, which is effectively mitigated through marker-making processes.

Both manual and computerized methods exhibit their respective advantages and disadvantages, and the selection between them depends on factors such as scale, complexity, and resource availability [5]. Ultimately, efficient marker making is pivotal for waste reduction, productivity enhancement, and the preservation of high-quality standards in manufacturing.

The marker holds immense importance in boosting garment production. Any inaccuracies or inefficiencies in marker making can severely disrupt the entire cutting process, leading to increased fabric wastage and decreased production output [5]. Through this research, efforts were made to identify factors contributing to marker efficiency. It is evident from this study that marker efficiency undergoes significant changes due to variations in garment size ratio and marker width.

References

1. Apparel Manufacturing Technology
2. K. Saha et al., *Impact of Sleeve Types on Marker Efficiency and Fabric Consumption* (2019)
3. S. Dominik, J. Henshall, J. O'Grady, K. Marshall, Factors influencing the efficiency of a marker-assisted introgression programme in Merino sheep. Genet. Select. Evol. **39**(5), 495–511 (2007). https://doi.org/10.1051/gse:2007017
4. Y. Xu, S. Thomassey, X. Zeng, An application of machine learning to marker prediction in garment industry, in *Proceedings of the 3rd International Conference on Applications of Intelligent Systems* (ACM, New York, NY, USA, Jan. 2020), pp. 1–5. https://doi.org/10.1145/3378184.3378219
5. E. Maqsood, H.-L. Chen, N. Abdulrahman, in *The Impacts of Different Woven Fabrics on Three Draping Techniques*, ed. by Elham Maqsood et al., p 147

Chapter 12
Impact of Digitalization on the Construction of Garments

Muhammad Babar Ramzan, Ayesha Kanwal, Syeda Maryam Hussain, Haleema Manzoor, and Hina Zehra

Abstract The garment industry is transitioning towards digitization due to the arrival of Industry 4.0. Digitization has benefitted the industry by streamlining and improving operations through the use of CAD and CAM. The garment manufacturing process has become more efficient and sustainable. However, many limitations need to be addressed such as the high investments and enhancing the skill of the workforce. Nevertheless, the seamless integration of these technologies has been proven beneficial as it led to improved quality of work. This chapter explores evolution along with the current tools and technologies integrated into today's industry. The advantages and limitations are also discussed in detail. Furthermore, the integration of modern technologies along with its impact on quality, sustainability concerns, and future trends is also discussed.

Keywords Digitization · Industry 4.0 · Sustainability · CAD · CAM · Automation · Training of workforce

---

M. B. Ramzan (✉) · A. Kanwal · S. M. Hussain · H. Manzoor · H. Zehra
Department of Clothing, School of Engineering and Technology, National Textile University Faisalabad, Faisalabad, Pakistan
e-mail: drbabar@ntu.edu.pk

S. M. Hussain
Department of Textile Design, Lahore College for Women University, Lahore, Pakistan

H. Zehra
Department of Textile & Clothing, National Textile University, Karachi Campus, Karachi, Pakistan

M. Q. Khan et al. (eds.), *Garment Sizing and Pattern Making*, SDGs and Textiles,
https://doi.org/10.1007/978-981-97-7683-2_12

Introduction to Digitization in the Garment Industry

The use of garments has existed since ancient times as it is considered one of life's necessities. Since the dawn of civilization, garments have not just been used for their functionality but also for culture, identity, and adornment. With the rise in urbanization, the domain of skilled craftsmen became the domain of globalized mass-production industries. The evolution of the garment industry is parallel to the evolution of society. The ever-growing garment industry has undergone four industrial revolutions. The first occurred in the eighteenth century as a result of mechanization and urbanization while the second took place in the twentieth century due to the introduction of new energy sources. The third Industrial Revolution occurred in the latter half of the twentieth century which was driven by the emergence of computers and electronics. The world is currently experiencing the fourth Industrial Revolution which is due to the advent of the Internet and digitization [1]. The aim of Industry 4.0 is to promote decentralization and interoperability, combining the virtual and the physical world to create dynamic production systems [2, 3] (Fig. 12.1).

With the globalization of the garment industry and the supply chain being scattered geographically, digitization has become a necessity. This has resulted in the investment in technologies that eliminate the need for physical proximity and allow for global communication [4]. Numerous technologies have been integrated into the garment industry that facilitate enhanced production efficiency due to the intense competition resulting from globalization. The combination of such technologies has resulted in reduced time and cost along with increased output giving the practitioner a competitive edge [1]. The result of this integration is not only subjective but also quantifiable. A study evaluated an investment project for digitalizing a technological process in a garment industry to decrease time loss and labour intensity. The

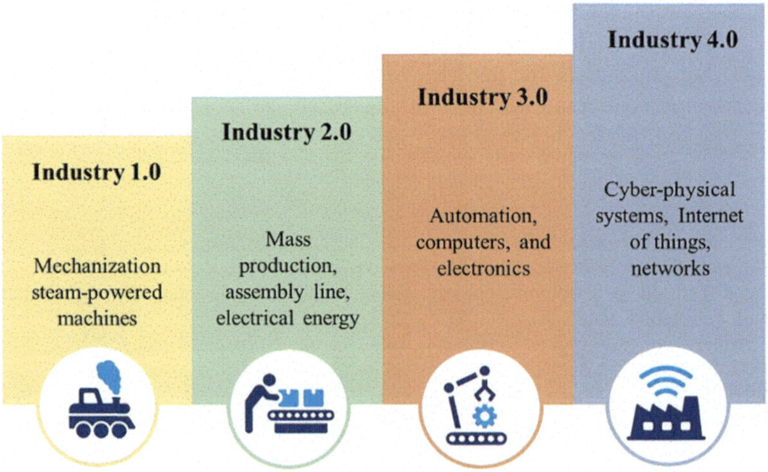

Fig. 12.1 Evolution of the Industrial Revolution from Industry 1 to 4.0

objective was to increase employees' available working hours to allocate more time for increased production. The results of this project were highly promising with a 30% increase in output due to improved productivity. Therefore, the hypothesis that digital technologies lead to increased efficiency became a fact [2].

In a world with continual industrial revolutions, industries must adapt to the latest technologies to succeed [1]. The garment industry has been using computer-aided design (CAD) and computer-aided manufacturing (CAM) for a considerable amount of time in advanced manufacturing processes; however, these technologies have significantly upgraded in recent years leading to digitization and automation [5]. Some of these transformations include technological advancements and revolutionized methods such as the use of 3D technologies, augmented and virtual reality (AR/VR), additive manufacturing, mass customization, made-to-measure (MTM), artificial intelligence (AI), big data, Internet of things (IoT), and cloud computing [3].

At the forefront is the use of 3D technologies that completely eradicate the need for physical proximity. The use of 3D technologies has contributed to dematerialization resulting in conserved resources and reduced environmental footprint. The garments can be designed from the initial concept to the final fitting in a complete 3D virtual platform. The use of these 3D technologies has already been adopted by market leaders like Nike and Victoria's Secret. Similarly, the use of AR/VR is also facilitating dematerialization by enabling consumers to interact with the products in real-world environments via smart or magic mirrors similar to DressX in many Japanese stores [3].

The emergence of MTM systems has also been revolutionary as it enables mass customization. With this new possibility of automating the entire garment manufacturing process, customized garments can be produced on a large scale. The MTM system seamlessly integrates with current CAD systems and allows for easy input of customer information, body measurements, and so on. All tools such as production schedules, plotters, cutters, and pattern design software are integrated into it [6]. Similarly, additive manufacturing has several benefits over subtractive techniques, one of which is the feasibility of on-demand and in-house production. Adidas has also made remarkable use of additive manufacturing in projects like FutureCraft which uses athletes' data to create customized midsoles, bringing mass customization one step closer. Furthermore, Clother in collaboration with H&M developed a way for people to vote on 3D-printed product prototypes. This enabled designers to assess market demand accurately [3]. Yet another great example is UnmadeOS® which creates a digital platform that enables the customization of products by connecting the front-end designing system with the Garment industrial knitting machines automatically. Similarly, Amazon has implemented an on-demand manufacturing system that will manufacture the garment once ordered. This configuration is designed to include a complete assembly line starting from printing and extending to cutting and sewing [5].

The role of IoT, big data, cloud computing, and AI in the digitization of the garment industry cannot be ignored. The use of AI has been multifaceted. It has been used to gain consumer loyalty by improving consumer experience by leveraging big data generated by the customer himself. This has also enabled the industry to generate much more accurate demand forecasts [3]. The use of big data, cloud computing, and AI by Macy's and Zolando is an applaudable example. They use a combination of these technologies for accurate demand and trend forecasts based on the shopping history of the consumer. It can also predict the right size for the customer based on the data of previous purchases [5]. When compared with advanced approaches, traditional forecasting methods are costly and highly inaccurate [4]. Furthermore, AI is being used throughout the industry for process improvement. For instance, artificial neural networks (ANN) are being used to identify areas of defect, malfunctions, and the reduction of waste [3].

Moreover, the use of IoT devices, big data, and cloud computing to make the garment industry transparent has been proven revolutionary. The traceability and transparency of the entire supply chain have significantly facilitated the industry's transition towards circularity and sustainability [1].

Technological advancements and digitization have facilitated the implementation of novel ideas in the clothing industry, such as high-speed sewing machines, CAD and CAM software, innovative cutting, fusing, and pressing techniques, and the use of robotics. There are two types of technological developments in the garment industry. Hardware technological advancements include automated sewing, automated identification, programmable production controllers, automated material handling, automated inspection systems, robotics, statistical process control, and so on. CAD, CAM, enterprise resource planning (ERP) software, statistical process control, production planning and inventory management software, and data management systems are examples of software technological advancements [7]. The utilization of the above-mentioned innovative technologies and procedures has resulted in several novel production, models, such as smart factories. Smart factories incorporate robotics, 3D technologies, and AR/VR technologies along with many others for hybrid production. These factories significantly reduce costs while dramatically improving the response rate, enhancing productivity and output. One great example is Hugo Boss which has been running the world's largest smart factory. Similarly, Adidas has been running a smart factory with significantly reduced human intervention [5]. At the current rate of technological advancements, we are not far from fully automated industries. For instance, several major corporations such as Levi's, Nike, and Adidas among many others are currently undergoing a digital transformation to digitize their processes [1].

Evolution of Garment Construction Techniques

When discussing the evolution, digitization, and automation of garment manufacturing, the use of advanced robotic machinery for product assembly comes to mind. However, the evolution begins at the fundamental level of product conceptualization and design. The world has transitioned from 2D product sketches to 3D computer-aided design. Instead of traditional paper, 3D virtual platforms incorporating 3D tools for sketching, draping, and styling are used. 2D to 3D synchronization has also been made possible which allows for modifications to be made instantaneously unlike the traditional method of redrawing and remaking patterns [5]. For example, tools such as VStitcher enable editing for fit improvement in a 3D virtual environment while simultaneously adjusting the pattern in 2D [6]. Similarly, Browzwear Lotta and Clo 3D are excellent examples of 3D garment design and visualization technologies. These technologies enable designers to test their designs on realistic body simulations without the need for fabric, paper patterns, and physical models [5]. CAD has also enabled the automatic gradation of patterns from a single block pattern using a pattern design system and a digitizer such as AccuScan and AccuMark from Gerber technology [6]. Recent researches indicate technological progress involving parametric design, AI, and 3D surface modelling. For instance, ANN can replicate the skill-based process of pattern-making by learning from input–output data and enhancing its descriptive accuracy and capacity. ANN models are being extensively trained with large amounts of experimental data, including body measurements and pattern coordinates to replicate various pattern-making processes [8]. Lectra Systèmes' is an important example of an organization dedicated towards technological innovations to offer complete CAD/CAM solutions for the garment industry [6].

Before the pattern-making stage, body measurements are required. This process has also been modernized in the twentieth century using 3D technologies [9]. Initially, the measurements were taken using a tape measure. Taking one's measurements was a time-consuming task and demanded skill for precision. 3D body scanners have significantly improved the process of taking body measurements by reducing the time required to mere seconds and increasing the accuracy tenfold. It also had a significant impact on the overall fit of the garment. Instead of having 10–20 measurement points, hundreds of thousands of measurements from every crevice of the human body have been made available. There has been significant progress in 3D scanning technologies with one iteration building on the one before it. For instance, one of the earliest 3D body scanners was the Loughborough anthropometric shadow scanner (LASS) which provided 3D measurements based on triangulation. The setup was minimal with minimal equipment requirements. The shortcomings of LASS were addressed by systems such as SYMCAD Optifit, and phase measuring profilometry (PMP) by providing instant data capture and configuration for improved accuracy and efficiency. Additionally, systems like Cyberware provided fast scanning of coloured data which led to enhanced accuracy. Further advancements included Intellifit's virtual fitting room (VFR) that incorporated cylindrical holographic imaging technology to provide quick and precise 3D body measurements, facilitating virtual fitting and personalized clothing suggestions. Every development of the 3D scanner has addressed the issue of speed, colour capture, user-friendly interface, and above all accuracy, in an ongoing effort for improved 3D body scanning. 3D to 2D conversion has also been made possible which automatically extracts patterns from the body scan. It is without a doubt that the development of the 3D scanner has paved the way for better fit, size, and visualization [10].

Once the measurements are taken and the patterns have been made, the materials need to be sourced from international suppliers which is a significant challenge

[5]. Before the globalization of the garment industry, all stakeholders were easily and physically accessible. However, following globalization, the materials needed to be sourced with a very long lead time. Improved communication technology has successfully resolved this issue. The new technologies allow for the real-time sharing of various types of data across all supply chain stakeholders. This has significantly reduced the lead time by ensuring that the right data is available at the right place at the right time [4].

The fabric spreading and cutting process has greatly evolved. At first, it was a time-consuming and laborious task to lay the fabric manually layer by layer. The cutting was also done manually on a limited number of layers, leading to inaccuracies and workplace injuries. There has been a notable improvement in the spreading and cutting process since the introduction of CAD technologies [6]. The machine can automatically lay the fabric while noting the fabric parameters such as length, width, and ply count. The sensors in the machine are present to align the fabric grain line. The fabric can be cut using a laser or water jet, eradicating the need for blades while reducing fraying and workplace injuries, increasing the smoothness of the cut and above all accuracy [7]. The automated spreading and cutting machines result in increased efficiency and reduced labour. The machines are equipped with integrated fabric fault management systems which identify substandard material resulting in improved quality [6]. Several models can automatically create a marker within their systems. This allows for reduced fabric waste and increased material utilization which aligns with the primary goal of digitization [8].

Primitive garments made of animal hide joined together provide us with a sense of ancient garment construction methods. Pre-historic sewing tools such as awls and needles can be found which have been made from bones, plants, wood, and bronze. Sewing needles made of steel were invented in the Middle Ages while the first sewing machine was introduced in 1750 by Weisenthal. Later a patent for a lock stitch wooden sewing machine was applied by Thomas Saint in 1790. By 1820, Barthelmy Thimmonnier built 80 wooden sewing machines, each capable of producing a different stitch. The wooden machines were replaced by metal machines by 1850. Since then the advancement in the sewing machine has been focused on speed and efficiency and in recent years automation. From semi-automated mechanized machines to profile sewing machines and then to fully automated sewing equipment, the sewing industry has been revolutionized [11]. The use of robotics in the sewing process has increased over the years. These robots are capable of handling and sewing the fabric. Sewbo by Zornow and Sewbot® offered by SoftWear are great examples of this advancement [5, 7]. These robots are capable of delivering 71% increased productivity along with precise and high-quality stitching [1]. Following the sewing stage of garment construction, sampling needs to be done. The 3D technology has eliminated the need for hundreds of physical samples. Despite great accuracy, creating a few physical samples is a necessity. Digital Twins play a significant role in connecting the virtual and the physical world in such instances [5].

Ensuring colour consistency among fabric components of a garment is an integral part of garment construction. Before the development of modern technology, colour requirements were communicated subjectively. Lab dips were commonly used and were extremely costly to produce. It needed to be sent for approval, and if rejected, they were returned and reproduced. This process was repeated until the lab dips were approved which usually took up to 5 tries and several weeks [4]. Metameric clothing was a huge concern at the time. With the advances in colorimetric, it became easier to achieve precise colour. The subjective description of the colour was replaced with objective data gained from object colour testing instruments. The garment industry has seen improvement in colour-matching techniques, including the use of spectrophotometers for accurate colour measurement and light booths for consistent lighting conditions. Colour matching and the numerical assessment of colour discrepancies have also become frequent uses for computer-aided techniques. The computer-aided colour-matching systems incorporate the skills of a colourist, shade banks, and shade cards. These methods forecast dyestuff mixes for spectral matching using mathematical models based on ideas like Beer's law and Kubelka–Munk theory. Today, spectrometers and colour inspection processes have made colour matching an easy and replica table process. With the advancement of colour-matching software and unlimited dyestuff, creativity in the garment industry is limitless. Computer-based colour-matching systems have become essential to the industry by providing economic advantages such as reduced raw material expenses, shorter lead times, and reduced production costs. In contrast to conventional techniques, which usually only achieve 30–80% success rates, modern technology such as Smartmatch® can achieve visually accepted first-shot results in up to 90% or more of predicted recipes [12].

The pressing process is extensively used throughout the garment manufacturing process to remove wrinkles and occasionally to create pleats and creases. It is the process that concludes the garment construction process. It soothes out the garment giving it an immaculate finish. Fundamentally the process of pressing involves heat, moisture and pressure. This simple equipment has evolved from a manual hot metal hand iron to a computer-controlled machinery. Historically, flat irons made of cast iron were used which were heated using a hot plate. These were modified to incorporate a thermostatically controlled electric heating element and later steam. On the other hand, the Hoffman press has been greatly used for high-volume garment production as an industrial counterpart of domestic steam iron. It is greatly valued in the garment industry due to its efficiency [13]. Although there are not many technological advancements in this phase of garment manufacturing, some new technologies have been introduced to the market such as pressing robots, jacket finishers, and shirt finishers. [7]. For instance, to gently iron the clothes to create an impeccable finish, a tunnel finisher is used. The garments are hung on a conveyor to be moved inside the steam cabinet and exposed to low-pressure steam with computer-controlled temperature and humidity. The speed of the conveyor and the time of steam exposure can be automated [13].

Following garment construction, inspection is necessary. Garment inspection has been done by skilled workers for decades which leads to inaccurate and inefficient detection of defects. Phycological and physical well-being played a crucial role in the accuracy of inspection. Automation tools with online automated systems using statistical, spectral, and model-based approaches have been introduced to automatically detect defects. The modelling tools extract information about the defect such as location, dimension, and severity [7]. The advancements in the garment industry will forever remain ongoing.

Digital Tools and Technologies in Garment Construction

The whole process of clothing production has three main categories, namelydesign, development, and production. Before the arrival of Industry 4.0, there were manual tools used for each step of production; for example, patterns were made by using manual drafting tools, and draping involved the placement of fabric against the mannequin or body form to create a garment. Garment and clothing designers faced many problems with the traditional methods, and to overcome all those problems, a magnetic evolution in every segment of garment construction has been brought through Industry 4.0 [14]. Various digital tools and technologies have been summarized here that have gained attention in the recent past.

3D Body Scanning

The very initial and important factor in creating a fitted garment is correct information about body dimensions. Taking measurements manually involved various drawbacks like time consumption and errors. However, 3D body scanners play a vital role in providing automated, efficient, and fast data for apparel manufacturing. In Loughborough, the very first body scanner to create a copy of the human body was developed [15]. The continuous revolution of technology has led to recent stereo-photographic fast scanners. The latest 3D scanners enable motion capture [15]. In simple words, the exterior of the human figure is copied by the 3D scanners. The combination of 3D scans and clothing patterns can be used for creating customized clothing. The further processing of these individually tailored digitized patterns can be done through grading and cutting automation [15].

As the clothing industry has evolved from mass production to made-to-measure to mass customization, consumers are inclined towards personalized clothing these days. Hence, the traditional pattern-making methods could not fulfil this demand resulting in a rapid revolution of pattern-making methods [16]. Digitizing table was one of the very initial technologies involved in pattern-making. It involved the use of special equipment that helped designers trace their hand-made patterns and convert them into soft form for further adjustment. Despite other drawbacks of this digitizing method, one of the most important aspects is time consumption due to which it got replaced by other digitizing processes. To fulfil the designer's requirements while providing accurate results, multiple programs are available in the market some of which are listed below.

- Modaris 3D Fit (Lectra),
- AccuMark VStitcher (Gerber),
- 3D Runway (Optitex),
- Virtual fashion (Reyes Infografica),
- Design concept (Lectra),

- Haute Couture 3D (PAD System),
- IT Simulator (Tukatech),
- Gerber AccuMark,
- Style 3D,
- Browzwear,
- Fitiquette [17].

Gerber Technologies

Gerber AccuMark is a complete suite of industry-leading software applications that help in transforming all steps from design to marker-making. Design creation, visualization of designs, and optimization of markers can be performed by using different software of AccuMark [17, 18]. Pattern design, development, and marker-making needs of the world's leading clothing manufacturers are dependent on Gerber AccuMark 2D. It has multiple automated processes, grading options, and industry-leading functionality of images that ensure its maximum efficiency. It brings the ideas quickly to life while increasing throughput and enhancing communication. Figure 12.2 shows different pieces of 2D patterns created in Gerber AccuMark 2D. Once the pattern has been created, different sizing variations can be created by just defining in software, and hence, all sizes will be covered immediately with a single command.

Gerber AccuMark 3D accelerates the whole process of apparel manufacturing and can be used as a source for bringing higher levels of collaboration and new efficiencies. It uses 2D pattern data to create realistic 3D renderings. It helps users in

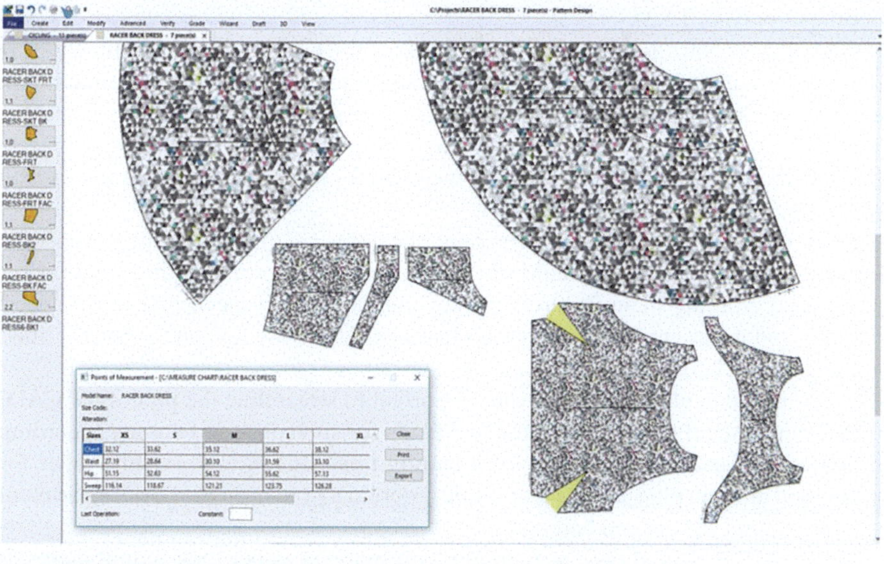

Fig. 12.2 Screen of AccuMark 2D sourced from [19]

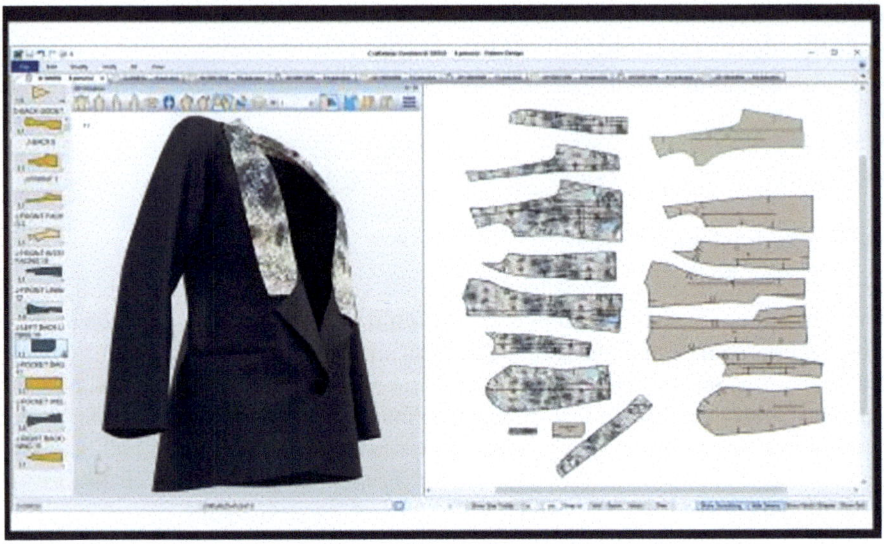

Fig. 12.3 AccuMark 3D interface sourced from [19]

visualization of their ideas and improve fit through virtual samples that are ready for production. The interface of Gerber AccuMark 3D is shown in Figure, where the 2D patterns are in the right window, and the 3D view of the same garment is displayed in the left window. At the same time, the individual pattern pieces are listed on the very left of the screen from where the user can select any individual piece for further modification (Fig. 12.3).

Gerber AccuNest generates cost and markers for production automatically. It is an automated application for nesting to accelerate the process and provide maximum material utilization by using vigorous algorithms. Figure 12.4 depicts a marker created in the AccuNest interface for single-ply. The length and width of fabric layers with several pieces adjusted are mentioned on the bar, allowing the user to modify them as per their requirements.

In the customized clothing environment, the items are made according to specific measurements and styles. AccuMark made-to-measure provides this solution for customized clothing manufacturers as they can use its advanced rule-based specifications and order creation automation features to provide fast and perfectly fitted clothing (Fig. 12.5).

To acquire real efficiency, it is very important to streamline the process of CAD-to-cut-room. AccuPlan creates optimized plans for spreading and cutting according to fabric rolls and the availability of a cutting table. It has a powerful tracker for production that can plan and track multiple orders at a time from the CAD department by using digital tickets and barcodes.

Gerber MP Plotter allows users to tint their markers as per their requirements. It has 2 or 4 print heads that maximize the throughput flexibility and speed. It adjusts

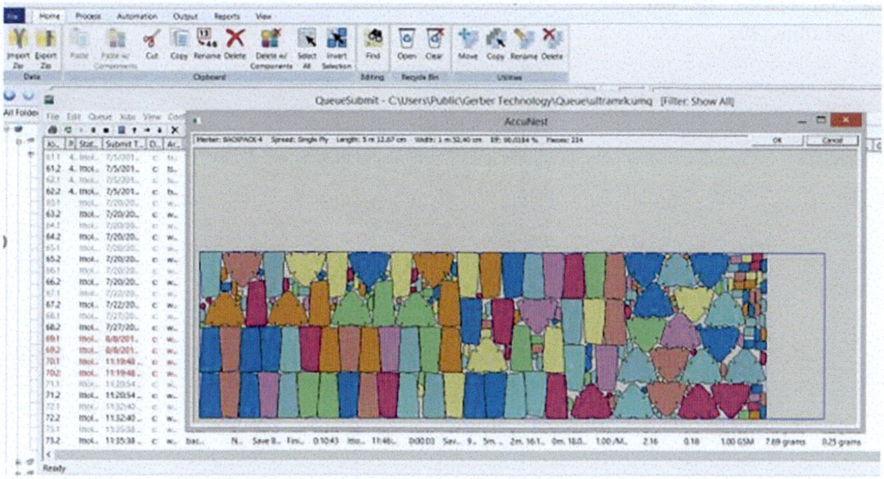

Fig. 12.4 AccuNest interface sourced from [19]

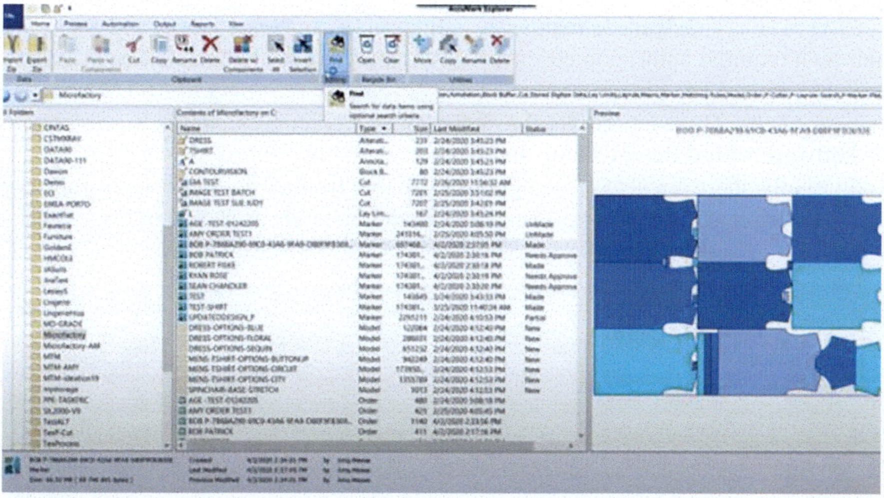

Fig. 12.5 AccuMark made-to-measure interface sourced from [19]

empty ink cartridges by using Techno-Ink cartridges, hence providing continued and uninterrupted plotting [17] (Fig. 12.6).

CLO 3D Software

CLO 3D is one of the most commonly used software among all other virtual fitting software available in the clothing market as it provides better performance in fabric simulation. It has user-friendly features and allows the user to import files from another CAD software. It has two windows on default, the right one displays 2D

Fig. 12.6 Gerber MP Plotter sourced from [19]

patterns, and the window on the left side shows a 3D view of that as shown in Fig. 12.7 [20]. CLO 3D has an array of options, and some of them have been mentioned below [21].

CLO 3D has an advanced feature that allows users to create virtual avatars as per their own required anthropometric data. Also, there are some default virtual avatars available in the CLO library that have been developed according to ASTM standard sizing systems. There are different avatars for males, females, and kids available in the software with different sizing and customization options as shown in Fig. 12.8. Additionally, there are also some options available for modifying the appearance of the avatar by changing its shoes, hair colour, and hairstyle as shown in Fig. 12.9.

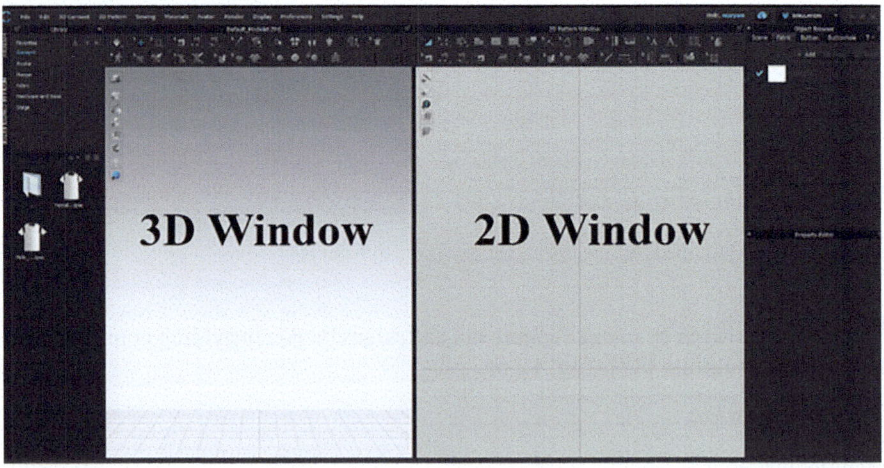

Fig. 12.7 CLO 3D interface showing both 2D and 3D window

Fig. 12.8 Virtual avatars available in CLO 3D sourced from [22]

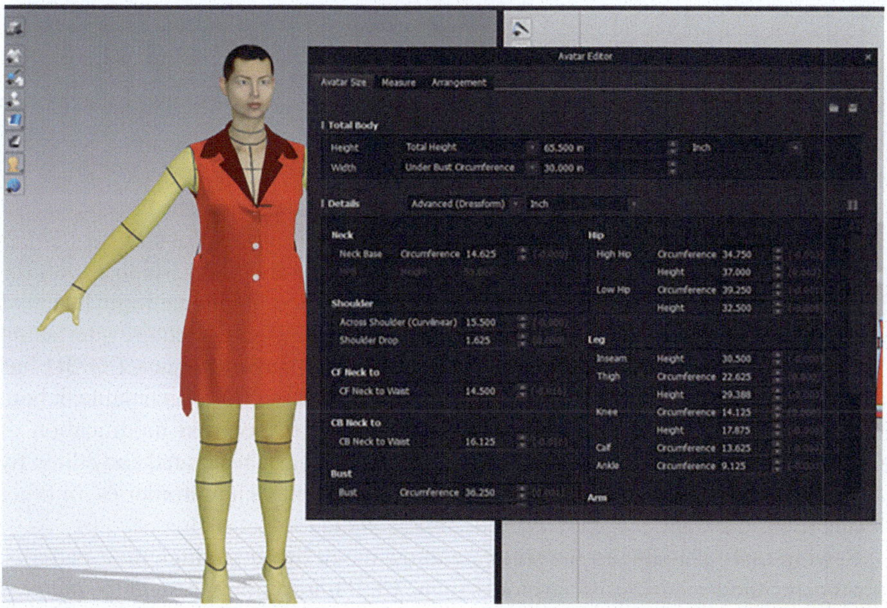

Fig. 12.9 CLO 3D avatar editor

The 3D window of CLO 3D has multiple tools for simulating the 2D pattern created in the 2D window. It enables designers to visualize their products before manufacturing and satisfy the demands of their customers without wasting time in physical sampling and reviews. There are multiple tools available for different editing and detailing options as shown in Fig. 12.10 [17].

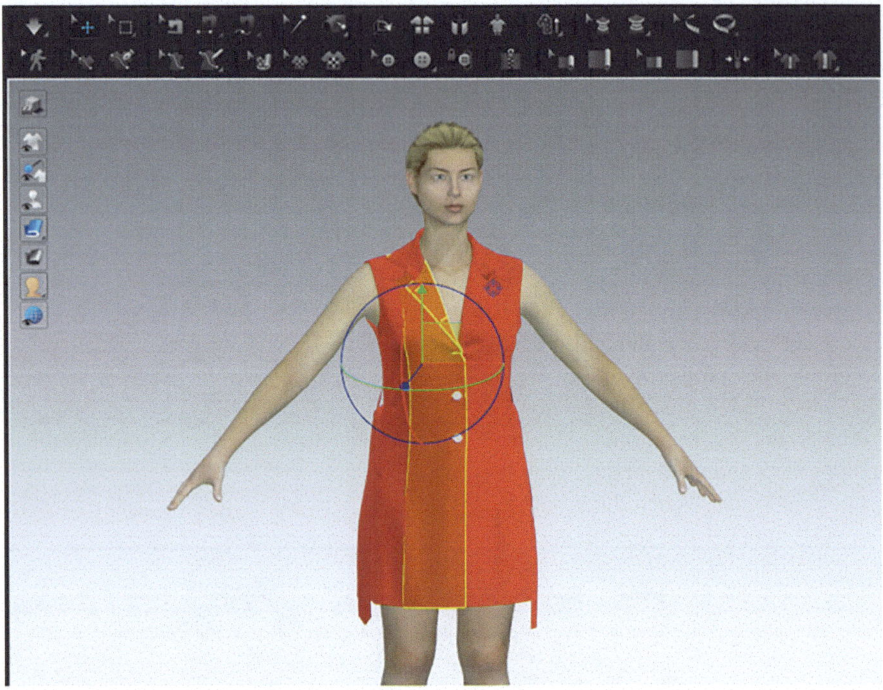

Fig. 12.10 3D simulation window with tools on top and left bars

Another incredible tool available in CLO 3D is its layering tool which allows the designer to create, modify, and visualize complex products having multiple layers of fabrics. Different clothing products like hoodies, coats, and jackets require interfacing fabrics which seemed a challenge for virtual clothing initially, but CLO 3D has successfully addressed these issues. There are multiple design tools available in both 3D and 2D windows of the software that allow the creation and modification of garment designs as per requirements. Different patterns can be created and edited by using these tools. These are useful in adding graphics, prints, embroideries, or other details to the garment.

Sewing tools available in the software allow the designer to stitch their patterns, and if using the already available blocks in the CLO library, they can edit the patterns by using the edit sewing tool from the 3D toolbar. The tool is available in both 2D and 3D window toolbars. When patterns have been made symmetrically, their seeing lines also move symmetrically. Visualizing printed garments or garments with some graphic elements virtually is also another incredible feature of this software. Designers can easily bring their prints developed in another CAD software to CLO 3D.

Fit evaluation can be conducted by using CLO software. It helps to review the instant impact of changes made to the pattern through immediate simulation. It checks the fit and silhouette sooner to improve the quality of designs [20]. There are three different fit maps available in CLO 3D, namely stress map, strain map, and fit map. These maps are available in the 3D window of the software. The stress map describes the strain in fabric which is caused due to the pressure applied on the garment by the human body. The strain map is described in the form of colours on the garment in the 3D window. The colour range is shown in Table 12.1. The areas of the garment with no strain appear in blue. The areas with normal strain appear in green colour. However, the areas with greater stretch will show the colours closer to red.

A stress map describes the pressure applied on the human body by the garment in a 3D window. This map shows the points using the colour range described in Table 12.2. The areas of the garment with no external pressure appear in blue. However, the areas of the garment with more pressure will turn closer to red.

Fit map describes the fit of the garment in terms of tight or cannot wear. Fit map highlights the spots on the garment where it gets tight, or the garment is not wearable. The range, percentage, colour, and opacity of colour have been described in Table 12.3.

CLO 3D enables the designers to display their designed clothing through animation on virtual models. It has a proper setup for creating virtual fashion walks where the garment can be viewed in motion during the catwalk. Different stages and runway options facilitate the user to show more creativity without having to create physical arrangements.

Garment Manufacturing Technologies

When it comes to the physical manufacturing of garments, there are various steps involved in the production of individual pieces of the garment. For example, the manufacturing of a pair of jeans requires many operations like pocket setting, loop

Table 12.1 Strain colour range

Range	Strain %	Colour	Opacity
Max	120.00		80%
Medium 1	117.14		80%
Medium 2	114.28		80%
Medium 3	111.42		80%
Medium 4	108.57		80%
Medium 5	105.71		80%
Medium 6	102.85		80%
Min	100.00		0%

Table 12.2 Stress map colour range

Range	Stress(kPa)	Colour	Opacity
Max	100.00		80%
Medium 1	85.71		80%
Medium 2	71.42		80%
Medium 3	57.14		80%
Medium 4	42.85		80%
Medium 5	28.57		80%
Medium 6	14.28		80%
Min	0.00		0%

Table 12.3 Fit map range

Range	%	Colour	Opacity
Max	100.00		80%
Medium 1	80.00		80%

placement, button attachment, etc. Similarly, while manufacturing a dress shirt there are multiple operations like pocket setting, sleeve placket arrangement, button attachment, collar attachment, etc. Some of the commonly used technologies in the garment industry have been described below.

Automatic Pocket Setter (for Jeans)

The process of setting pockets that were traditionally done manually has now been replaced by an automated pocket setter machine. This machine has a sewing area of 250 * 250 mm. It also has a stacker with a capacity of 64 pieces that can complete an operation series of pocket-setting procedures. There is space within the main machine body to store the stacker inside it. This machine replaces 8 workers with just one as this whole system can be operated by a single operator [20].

Automated Sleeve Placket Attachment under Computer Control

The positions of the front and placket of the shirt can be done accurately by using laser lights of the system. The cycle of sewing starts automatically, once the position of the garment piece is changed. Garment gets piled after sewing and cutting of

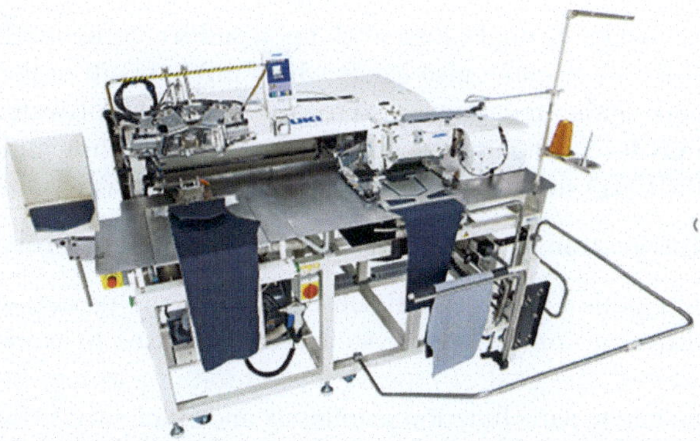

Fig. 12.11 Automatic pocket setter machine by JUKI sourced from [24]

thread. It uses a belt feed system for putting the garment pieces. Its control panel is a touch LCD where the parameters for work and patterns can be programmed [23]. The machine can be seen in Fig. 12.11.

Advantages of Digitization in Garment Construction

Generally, digitization is the process of converting a traditional business into a digital business while generating revenue and value-added opportunities. It has completely transformed each step of garment manufacturing. Garment industries have observed multiple benefits of shifting towards digitalization. For instance, the design is produced with high accuracy, a smaller number of operators are required, the chance of error is reduced, and designs and samples can be revised and remade. The average lead time had been reduced by 2–9 weeks while the sample quantity needed had been reduced to 50%. Designers can alter designs in less time and at no cost. Productivity has increased substantially. Designs produced by using CAD systems can be saved easily and used again in the future whenever needed. Virtually rendered designs provide a realistic view of the design with multiple options while the creation of template and database feasibility is provided. Furthermore, it facilitates product life cycle management [14, 17].

Challenges and Limitations of Digitization

Designers using different CAD software face big challenges and limitations. To achieve the same design work while using different CAD software that is based on different functionality and performances can be a big challenge. The recognition of different CAD software in a short time is a difficult task. For example, it is difficult to import CAD models in software that has a linked modelling platform. Moreover, some of the products fail to meet people's needs as the designers relying completely on CAD software lack problem-solving ability. The product's virtual appearance diverts the focus of the designer from the product's utilization in real life. Additionally, CAD systems require high-quality hardware for computers and consume more power which causes an increase in cost. The proper maintenance of CAD devices can be challenging on the available resources. If not saved on the cloud, the designer might lose work upon sudden computer breakdown. The possibility of software hacking also exists which might result in the loss of all work. Continuous and regular updating is also required by CAD software. Furthermore, first-time users may face difficulty in learning the complex technicalities of the system. For utilizing the CAD software properly, detailed training is required [17].

Integration of Digital Design and Production Processes

Integration, in a general sense, refers to the process of combining or coordinating different components, systems, or processes to work together as an incorporated whole. The goal of process integration in garment manufacturing is to ensure that disparate parts function seamlessly and efficiently, to improve performance, functionality, and overall coherence in processes. The production of garments with industry is complex, and integration between stakeholders and processes plays a crucial role in creating seamless and interconnected solutions. The integration of digital design and digital production tools creates a continuous and efficient workflow, from the conceptualization of garments to their production and distribution. This integration not only enhances precision and speed but also contributes to sustainability and innovation in the apparel-making process [25]. Figure 12.12 is a process diagram presenting how garment production processes are integrated to give the best possible products.

From current research on distinctive between textile clusters and integration of garment production systems, the matter for predicting future advances in technology fields and business customer sectors is not clear enough. These limitations open new opportunities to implement Industry 4.0, as some stakeholders are more interested in digitally controlled production processes. After a systemic understanding of process efficiency in an organization, high-quality garment production and quality management will synchronize the digitalization of the system, and the processes must be upgraded to each company's production requirements. The ambitious smart factory concept includes intelligent manufacture control, constant self-optimization

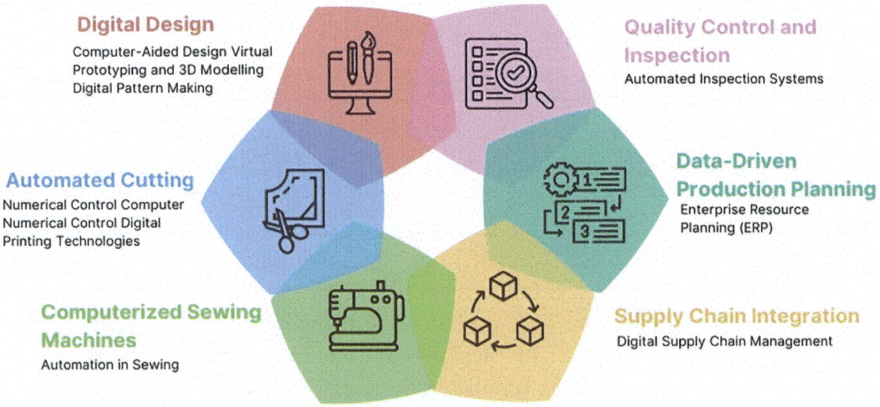

Fig. 12.12 Integration of garment production processes

real-time production data processing (big data analytics and artificial intelligence), the Industrial IoT, and interconnected machine-to-machine communication through digital integration within production by machine reprogramming creating fully flexible production processes that open solutions to economically possible massive customization [26].

The cloud-based digital communication and collaboration platforms help the designers, manufacturers, and other stakeholders collaborate in real time. This enhances communication, reduces errors, and accelerates decision-making. The importance of centralized digital design libraries is undeniable these libraries store and manage design assets, patterns, and specifications. This ensures consistency across the design and production processes. Computer numerical control (CNC) cutting machines use digital patterns to automate the cutting of fabric ensure precision, reduce material waste, and accelerate the cutting. Digital printing technologies help in digital textile printing for high-resolution and detailed prints directly onto fabric. This eliminates the need for separate printing processes and contributes to sustainability by reducing water usage and ink wastage. The automation of sewing processes is achieved by using computerized sewing machines which use digital instructions to automate the sewing process and guarantee consistent stitching quality, reduce manual labour, and enhance production efficiency. Quality control and automated inspection systems are employed for digital quality control. These systems detect defects and ensure that garments meet quality standards before progressing further in the production process. The efficient use of resources and reduction in production lead times are managed in data-driven production planning software [27].

Enterprise resource planning (ERP) systems are digital tools used for data-driven production planning, optimizing resource allocation, and scheduling. This ensures efficient use of resources and reduces production lead times. The supply chain integration through digital supply chain management tools across the supply chain facilitates real-time visibility into inventory levels, production status, and shipping information. This transparency improves collaboration among suppliers, manufacturers, and distributors. Digital tools enable real-time monitoring of production processes. Analytics provide insights into production performance, defect rates, and overall efficiency, allowing for continuous improvement. Digital design and production tools support feedback loops, allowing for continuous improvement based on insights gained from previous projects. This contributes to iterative design and enhanced production efficiency. Integration of digital tools supports the adoption of sustainable technologies, such as energy-efficient machines, eco-friendly materials, and practices that minimize environmental impact. Blockchain technology can be integrated for enhanced transparency and traceability, ensuring ethical sourcing and authenticity in the supply chain [28].

Impact on Quality and Precision in Garment Construction

The incorporation of these digital tools not only enhances the efficiency of the garment design and construction process but also contributes to sustainability by minimizing material waste, improving precision, and optimizing the supply chain. As technology continues to advance, the fashion industry is likely to see further innovations in digital design and garment construction processes. Quality control (QC) is an important part of the production process. Inspectors examine garments for defects, ensuring that they meet the specified quality standards [29].

Table 12.4 is a representation of the comparison between improvements that lead to good quality in garment manufacturing processes through digital tools and traditional methods. This may involve visual inspections, measurements, and functional tests. The integration of digital design and production processes has revolutionized the garment industry, streamlining workflows, enhancing creativity, and improving efficiency throughout the supply chain. This convergence of technology and traditional practices has paved the way for more sustainable and customer-centric approaches to garment manufacturing [30].

Ensuring the quality of garments is crucial for customer satisfaction, brand reputation, and compliance with industry standards. The digitization of quality-checking processes in garment manufacturing contributes to a more streamlined, efficient, and error-free production environment. It not only ensures the production of high-quality garments but also supports sustainability, cost-effectiveness, and customer satisfaction. Manufacturers benefit from improved processes, reduced waste, and the ability to adapt to evolving industry demands. The quality of good garments is associated with many factors according to garment end usage for example fabric quality, colour fastness, seam finishing, accuracy in size, pattern matching, right labelling, care instructions, and brand information. The advancements in technology, such as the adoption of automation and digitalization, continue to impact production processes in garment manufacturing, leading to increased efficiency and precision in various stages of production. The quality and precision in garment manufacturing are not only critical for meeting customer expectations but also for sustaining a successful and competitive business. The impact extends throughout the supply chain, from efficient production processes to customer satisfaction, and contributes to the long-term success and reputation of garment manufacturers on an industrial scale.

Currently, the largest obstacle to garment design and manufacturing is the evaluation of garment fit. The forecast accuracy of the fit assessment model based on Back Propagation Artificial Neural Networks (BP-ANNs) with 93% accuracy can assist with the implementation of a garment fit prediction model utilizing data learning technology based on Artificial Neural Networks. Related to the 50% forecast accuracy of the traditional method, a new method has clear advantages. This technology can be applied to the process of garment design and manufacturing to expand work effectiveness [31]. Previous research on digitalization has highlighted the impact of the transformation on the business models of an organization to advance new value networks or a completely innovative business model grounded on smart technologies [1].

Sustainability Considerations in Digitized Garment Production

Sustainability is a broad and multifaceted concept that encompasses practices, strategies, and principles aimed at meeting the needs of the present without compromising the ability of future generations to meet their own needs. It involves the responsible use of resources, consideration of environmental impact, social responsibility,

Table 12.4 Comparison of traditional and digital methods for quality improvement of garments

Aspect of improvement	Digital tools in garment manufacturing	Traditional methods in garment manufacturing
Design and pattern-making	CAD software for precise digital pattern creation 3D modelling for virtual prototyping	Manual pattern drafting is often prone to errors Physical prototypes for design validation
Material cutting	CNC cutting machines for precise and automated cutting Automatic nesting software for optimal fabric utilization	Manual cutting with scissors or rotary cutters Manual pattern placement leads to material wastage
Sewing processes	Computerized sewing machines for automated stitching Collaborative robots (cobots) for assistance in sewing tasks Programmable stitching for customization	Manual sewing with traditional machines Limited automation leads to slower production
Quality control	Computer vision for automated defect detection Real-time monitoring and analytics	Visual inspection for defects is often subjective Post-production quality checks
Supply chain management	Digital systems for real-time data exchange. Cloud-based collaboration platforms for seamless communication	Manual communication and documentation Lack of real-time visibility in the supply chain
Flexibility and adaptability	Agile production with quick response to design changes Remote monitoring and control of production processes	Time-consuming setup for design changes Limited flexibility in adapting to market trends
Resource utilization	Data analytics for optimizing resource allocation Barcode/RFID technology for tracking and traceability	Limited visibility into resource usage Manual tracking and record-keeping
Environmental impact	Digital tools for optimized material usage, reducing waste Sustainable practices through digital printing and cutting technologies	Higher material wastage in manual cutting Traditional processes may contribute to environmental concerns

and economic viability. Sustainability seeks to create a balance between economic development, environmental stewardship, and social equity. Sustainable development goals (SDGs) with digital construction methods involve interconnection and mutual reinforcement between SDGs and digital construction methods. Figure 12.13 is a visual representation of SDGs. By incorporating these sustainability considerations into digitized garment production, the fashion industry can move towards more environmentally friendly, socially responsible, and economically viable practices. It

Fig. 12.13 Visual representation of SDGs

requires a holistic approach, collaboration across the supply chain, and a commitment to continuous improvement [32].

Modern trends could substantially improve the working conditions in modern textile factories. The digitized textile firm offers more sophisticated tasks to its workforce, creative interaction with other value chain partners, options to contribute to technological advancements, and permeable career paths for many employees. The environmental impacts of digital transformation in manufacturing, especially in the optimization of resource and energy use through digital technologies, are proven through research. Sustainability in garment construction is increasingly recognized as crucial in the fashion industry due to its significant impact on various stakeholders, the environment, and society as a whole [1].

Traditional garment manufacturing processes often involve significant resource consumption, including water, energy, and raw materials. These sustainable practices aim to minimize these impacts, promoting the conservation of natural resources. Traditional garment manufacturing processes often involve significant resource consumption that leads to climate change and increased carbon footprint. Sustainable garment construction aims to reduce the carbon footprint associated with the fashion industry. This includes minimizing greenhouse gas emissions from production processes, transportation, and end-of-life disposal by adopting energy-efficient processes and utilizing eco-friendly materials to mitigate climate change impacts. Traditional garment production often involves high water usage, chemical inputs, and energy consumption. Sustainability focuses on diminishing these impacts by approving eco-friendly materials, plummeting water usage, and optimizing energy yield. Due to the amplified global consciousness to save the planet Earth, many foremost fashion brands have the attentive urge to decrease the pollution produced by the manufacture of garments [33].

Conventional textile manufacturing can involve the use of harmful chemicals and dyes causing environmental pollution. Sustainable practices focus on reducing or eliminating the use of toxic substances, leading to less environmental pollution. The garment industry is notorious for generating substantial textile waste. Sustainable practices aim to reduce waste through measures such as recycling, upcycling, and designing for longevity. This helps in minimizing the environmental burden associated with textile disposal. The garment industry generates substantial textile waste, both during production and post-consumer use. Sustainable practices, such as recycling materials, upcycling used clothing, and designing for durability to minimize the environmental impact of textile waste [34].

Promoting reuse and recycling is the aim of sustainable garment construction by embracing the principles of the circular economy, where products are designed for longevity and are recyclable or biodegradable. This approach minimizes the end-of-life impact of garments and promotes a closed-loop system. The goal of sustainable clothing manufacturing is to encourage reuse and recycling by adopting the circular economy's tenets, which call for long-lasting, recyclable, or biodegradable products. This strategy encourages a closed-loop system and reduces the impact of clothing at the end of its useful life. Thus, supporting long-lasting product design and encouraging material reuse and recycling [35].

Consumers have become more environmentally conscious, there is an increasing demand for sustainable and ethically produced garments. Brands that prioritize sustainability in construction are more likely to resonate with consumers who value eco-friendly practices. For meeting consumer expectations, brands that prioritize sustainability in garment manufacturing are likely to have more business. Conscious consumers are seeking products that align with their values, including environmental and social responsibility [36, 37].

Brands that prioritize sustainability in garment construction can build trust and loyalty among consumers. A positive brand image, associated with responsible and ethical practices, contributes to long-term success and competitiveness in the market. Brands that demonstrate a commitment to sustainability in garment manufacturing can build trust with consumers. Thus, a positive brand image associated with responsible practices can lead to customer loyalty and long-term success [38].

Sustainability in garment construction is integral to addressing environmental, social, and economic challenges associated with the fashion industry. By adopting sustainable practices, garment manufacturers contribute to a more responsible and resilient industry that benefits both current and future generations [39]. Sustainability in garment construction extends to ethical labour practices by ensuring fair working conditions. This includes ensuring fair wages, safe working conditions, and adherence to labour rights. Brands committed to sustainability often prioritize transparent and ethical supply chains. Ethical labour practices are integral to the social responsibility aspect of sustainability [40].

The pursuit of sustainability in garment construction encourages innovation in materials, production processes, and business models. Collaboration among stakeholders, including designers, manufacturers, and suppliers, fosters the development of more sustainable practices which drive innovation in the industry [41].

Governments and international organizations are increasingly introducing regulations and standards to promote sustainability in the fashion industry. Adhering to these regulations is essential for the long-term viability of garment brands. These standards are essential for compliance and reputation management [42].

Incorporation of Digital Tools in Garment Production Processes with Sustainable Practice

The importance of sustainability in garment manufacturing has grown significantly in recent years due to the recognition of environmental, social, and economic challenges associated with this industry. So, sustainability in garment manufacturing is essential for addressing environmental and social issues. It involves a comprehensive approach that considers the entire lifecycle of a garment, from design and production to use and disposal. Embracing sustainable practices is not only ethically responsible but also essential for the long-term success and resilience of the fashion industry. Sustainability considerations in digitized garment production involve adopting practices and technologies. In the context of digitized garment production, the advantages of digital tools in enhancing efficiency, precision, and sustainability in garment manufacturing processes are undeniable. The relationship between sustainability and garment manufacturing through digitization in processes is significant. Digitization in garment manufacturing involves the use of digital technologies, automation, and data-driven approaches to streamline various stages of the production cycle. This corporation of digitization and sustainability brings about several positive impacts like it enhances efficiency, reducing environmental impact, and supporting ethical practices throughout the production cycle. Fig. 12.14 presents a contribution to achieving sustainable goals. It allows the industry to move towards more responsible and sustainable practices, meeting the demands of environmentally conscious consumers and addressing global sustainability challenges [43].

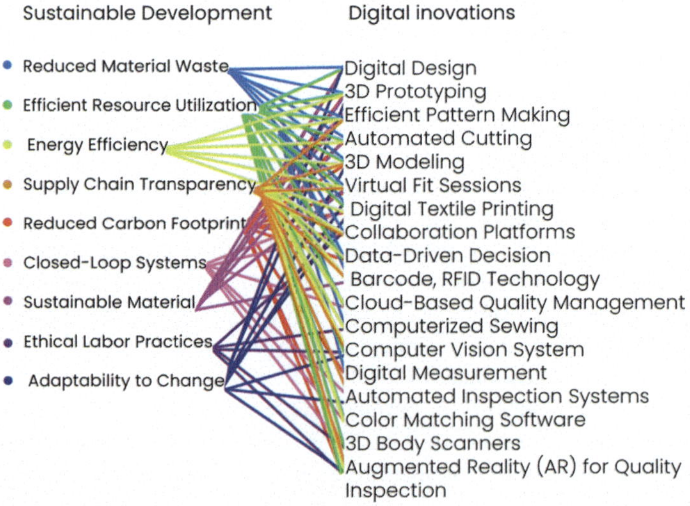

Fig. 12.14 Corporation of digitization and sustainability

Training and Skill Development for Digital Garment Construction

The garment and apparel industries are labour-intensive and were the key drivers for industrialization globally [44]. The garment industry is evolving rapidly and producing several clothing and fashion products according to style and seasons. The manufacturers constantly face challenges in such circumstances. They are anticipated to produce a cheaper, faster, and more sustainable supply chain while effortlessly integrating new technologies into methods of manufacturing [45]. The emphasis on automation under the fourth Industrial Revolution vividly changes how the apparel industry operates [46]. The garment manufacturing industry has a historical background of adopting technologies and automation. With this evolution in the industrial sector, companies and brands need to be more innovative to meet the market requirements. Usually, industrial policies suggest high-value products. However, these high-value products require digitization and a skilled workforce to operate these advancements [45]. Technical skills are often lacking in the workforce in developing countries for advanced production methods. According to numbers, more than 60% of vendors find it difficult to fill skilled workforce positions [46].

Importance of Training and Skill Development

For the enhancement of abilities and dexterity of employees to make them able to perform efficiently and productively, training and skill development are crucial. Training improves the skills and knowledge of the employees and helps them to understand the particular skills that are required for the enhanced efficiency of the employees and production. It is a way to manage and improve the quality of manpower in an industry either new or existing. Training helps in motivating the new joiners in the organization. It increases their efficiency to sustain the high standards of customers and product quality. It is also important to introduce new advancements and technologies. When the workforce is familiar with new tools and technologies, they will support the organization by adapting them. Familiarity with technological advancements will improve productivity in a positive manner which ultimately increases the profit and efficiency of an organization [47].

In this rapidly changing world of digitalization, training and skill development are crucial for success. It offers opportunities for the workers and enhances production methods as well. With the passage of time bombarding technologies, it is vital to align the workforce with the required changes, and training is one of the best ways to refine employees as per the need [48, 49]. The workforce is the core factor that distinguishes one industry from another. For industrial sustainability in the dynamic market, the quality of the workforce is essential [47]. It is worthless to talk about improvement without training and educating human resources. Therefore, organizations need to train their employees to increase efficiency, productivity, and creativity over time as shown in Fig. 12.15 [50].

Impact of Industry 4.0 on Training and Development of Manpower

The human workforce is being substituted with automation cause the faster global advancements and digitization. The nature and scope of work around industries

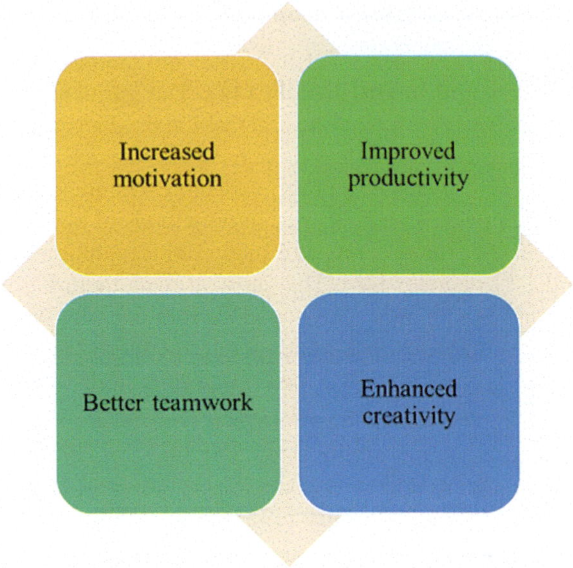

Fig. 12.15 General benefits of training and skill development

changed by this development as well. The technologies will reshape future jobs, such as manpower in repetitive jobs may be replaced with robots. According to the literature, approximately 60% of workers will lose their jobs shortly, around 2027 because of advancements in technology. However, according to the World Economic Forum report, new job requirements and skills will be required by Industry 4.0 which will change the way of working. With these rapid advancements, the systems will get digitalized much more efficiently. In this context, it can be said that the jobs will get automated sooner or later. New jobs and occupations for the workforce will emerge as a human–machine developed era. The change in the business models due to digitization will change the job market by introducing new jobs [47].

Skills Required for the Digital Garment Construction

The digital garment construction industry requires a blend of skills to sustain which includes traditional knowledge and proficiency in digital tools and technologies as well. The required skills can be categorized as soft skills, technical skills, and domain-specific skills.

In the contemporary industry, the enhancement of soft skills is crucial for an attractive career and exceptionally competitive global apparel market. Communicational skills consist of verbal, non-verbal, and listening skills. Verbal skills are about what and how we say, and non-verbal communication is related to the tone of voice or body language. Listening skill is related to interpreting verbal and non-verbal communication by others [51]. In the world of automation and artificial intelligence, the industrial sectors are focusing on boosting information communication among departments. Effective communication practices in industries can help in sustaining the neck-throat competition in this contemporary era. To run a smooth production, the involved workforce must be familiar with the constraints at each level. Poor/ weak communication at the workplace may have a negative effect on the output because of dissatisfaction with the job which leads to a lack of understanding and low productivity [52].

The problem-solving skill is crucial for digital garment construction for numerous reasons. Digitization in the garment manufacturing industry involves working with complex tools, technologies, and software. The workforce may encounter challenges and limitations while working with 3D modelling programs, CAD software, or virtual prototyping. Problem-solving skills help the workforce tackle the limitations and challenges efficiently. The rapidly evolving nature of technology in the digital garment construction industry makes it essential for the workforce to be adaptable to change. Designers must be flexible to new tools and technologies to adjust their workflow diversely and effectively join forces with global markets. Embracing adaptability led the designers to stay current with technological advancements and deliver innovations in the fashion industry [53].

Domain-specific skills are the ones that are particularly required in that specific area. They are beyond the general and technical skills such as specified software or communication and problem-solving skills and require a delve specialized knowledge to excel in a specific area. Digitization in the garments manufacturing industry demands several skills to work efficiently in the global market such as knowledge about garment construction, familiarity with textiles, knowledge of industrial standards, sustainability awareness, and expertise in the design and manufacturing of garments [53].

Computer-aided design (CAD) software is important in the digital garment construction industry because of numbered reasons. CAD software is considered one of the most vital tools in the fashion and garment industry. It has been used for pattern-making, grading techniques, and marker-marking processes. It enables the workforce/designers to design accessible and accurate digital patterns. Proficiency and understanding of CAD software encourage the workforce to create high-quality, efficient, and customized garments to keep pace with the global market [54]. Other essential tools are digital prototypes that are frequently used to streamline the designing and production process of garment manufacturing. Although e-prototyping and digital designing yield greater sustainability and efficiency in the apparel industry, these practices are not fully mature. For commercial solutions, it is essential and better to know these available 3D solutions such as AccuMark3D (Gerber), CLO3D, Optitex3D, Style 3D, VStitcher, or Tukatech. 3D modelling software and virtual prototyping are decisive in digital garment construction. They allow to capture of the actual and accurate measurements and body shapes which facilitates the personalization and custom-fit garments. In the digital garment manufacturing industry, digital textile design software such as Adobe Photoshop and Adobe Illustrator are essential to learn and have a pivotal role in enhancing design quality. By understanding digital tools and technologies, the workforce especially designers can incorporate detailed, precise, and accurate patterns and prints with unparalleled flexibility and endless creativity. Furthermore, it facilitates customization and allows designers to customize and personalize garments swiftly. This sleek process not only accelerates the process but also makes high-quality production by minimizing time as well. Moreover, the use of digital textile design tools with other tools and technologies such as virtual prototyping and pattern-making enables the designers to envision and assess designs in the background of the final garment. However, the skill development of such tools and software enhances the whole design process and also contributes to sustainability by reducing waste and minimizing harmful environmental impacts [53].

Different Methods of Training

The workforce in the digital garment manufacturing industries requires the skills to handle 3D design software and digitized and automated production systems. While traditional training systems are beneficial, online training platforms offer numbered advantages. Online training sessions break down the geographical barriers and allow global access. Online training programs are low-cost and scalable which makes them ideal for individuals as well as for industries. Such programs empower the workforce in the digital garment manufacturing industry to enhance its dynamic landscape and ensure a skilled and competitive workforce [55].

The Internet of things (IoT) is reshaping the garment manufacturing industry swiftly. The workforce that attends the seminars and training related to IOT can learn the basics of electronics, sensors, and digitized tools and technologies. Companies or organizations can also offer on-the-job training and skill development programs in garment manufacturing industries to provide mentorship to their workforce [55].

Future Trends and Innovation in Digitization of Garment Construction

Reports indicate that in the near future 3D design, digital printing, virtual prototyping, and RFID will get familiar, demanding, and highly impact the global garment manufacturing and fashion markets. Several brands have already started implementing such technologies in their businesses. Such as H&M and Zara are overruling and live streamlining their runways and catwalks and using "see now buy now" models for this generation instead of using traditional methods to be implemented in their supply chain [56].

CLO3D is in the lead in the garments manufacturing and fashion industry. The software technology is continuously evolving and will nearly incorporate AI-powered suggestions for design, advanced algorithms for fabric simulation, and other software and hardware device integration. Also, it will handle the geographical constraints and limitations by creating digital showrooms and fashion walks [57]. Implementation of augmented reality (AR) and virtual reality (VR) techniques allows customers to virtually try on apparel articles, by experiencing different textures, drapes, and colour contrast and gives a 360-degree view in virtual reality. It enhances the online shopping experience and reduces returns. Many leading brands are in the process of incorporating this technology [58].

The garment manufacturing industry is on the edge of the robotics revolution. This technology is being started by the adoption of robotic systems for different tasks like fabric cutting, sewing, and finishing, increasing production speed, consistency, and precision while reducing labour costs. It will improve the new ways of automated garment manufacturing. The rise of globalization and digitalization is continuously increasing the need for flexible manufacturing processes. By integrating robotics into the garments manufacturing industry, manufacturers can speed up production with precision and accuracy. Companies are investing in such technologies to meet global needs and to see the right results. Robotics technologies are working and integrating and, in the future, will modify almost every department of garment manufacturing from fabric testing to inventory management including printing, cutting, sewing, and assembly processes as well [59].

In this era, customers are becoming more aware and demanding of ethical practices. Because of that traceability and transparent supply chains become crucial factors for garment manufacturing industries. By using the blockchain and transparent supply chain systems, companies and customers are getting end-to-end visibility which ensures accountable production practices and improved customer trust. Traceability systems such as RFID tags, barcodes, and QR codes allow companies to track and monitor the raw materials to finished products movement [60].

One of the major trends that is reshaping the global apparel and fashion industry is sustainability. Industry 4.0 is embracing sustainable practices by incorporating digitization and technologies in the garments manufacturing industry. Technologies, such as 3D printing, aided in the sustainable production of personalized clothing that will fit them specifically. Approximately 10% of greenhouse emissions come from the textile and clothing sectors. Therefore, practising digitized and sustainable practices in the garment manufacturing industry will filter the production methods and impact the longevity of whole economies. Several currently using technologies are helping the sustainable clothing sector. However, these innovations require time, money, and effort as well to reach their full potential [61].

The role of artificial intelligence in design optimization is keenly helpful in the garments manufacturing industry. AI systems predict the demands, cope with the supply chain, and polish the quality control where needed. In the coming time, the incorporation of AI algorithms to evaluate and analyse consumer data, preferences, and trends to inform design decisions, predict demand, and optimize product offers for maximum market appeal [62].

Apart from these, the expansion of online platforms will facilitate collaboration between designers, manufacturers, and consumers in the garment manufacturing design process, supporting creativity, innovation, and co-creation. These trends and innovations collectively signify a shift towards a more technologically advanced, sustainable, and consumer-centric approach to garment construction, driven by the union of digital innovation and the evolving needs and expectations of the fashion industry and its consumers [63].

References

1. A. Kaplanidou, Digitalization in the Apparel Manufacturing Process. (2018), [Online]. Available: https://studenttheses.uu.nl/handle/20.500.12932/39255. Accessed 26 Feb 2024

2. T. Y. Kudryavtseva, K. S. Kozhina, G. K. Prause, E. O. Olaniyi, Assessment of economic efficiency, effects and risks of digitalization projects of garment industry in Russia. J. Appl. Econ. Res. **22**(1), 72–98 (2023). https://doi.org/10.15826/vestnik.2023.22.1.004
3. R. S. Kanwal Ayesha and Nazeer, Digitalization in the textile sector for circularity, in *Circularity in Textiles*, ed. by S. and N. Y. and H. M. Batool Syeda Rubab and Ahmad (Springer Nature Switzerland, Cham, 2023), pp. 213–231. https://doi.org/10.1007/978-3-031-49479-6_8
4. D. Tyler, Advances in apparel product development, in *Advances in Apparel Production* (Elsevier Inc., 2008), pp. 157–177. https://doi.org/10.1533/9781845694463.2.157
5. R. Pal, A. Jayarathne, Digitalization in the textiles and clothing sector, in *The Digital Supply Chain* (Elsevier, 2022), pp. 255–271. https://doi.org/10.1016/B978-0-323-91614-1.00015-0
6. T. Bond, Computerised pattern making in garment production, in *Advances in Apparel Production* (2008), pp. 140–153. https://doi.org/10.1533/9781845694463.1.140
7. R. Nayak, R. Padhye, Introduction to automation in garment manufacturing, in *Automation in Garment Manufacturing* (Elsevier, 2017), pp. 1–27. https://doi.org/10.1016/B978-0-08-101211-6.00001-X
8. Y. Xiu, Z.-K. Wan, *A Survey on Pattern-Making Technologies in Garment CAD*
9. C. Fairhurst, Advances in Apparel Production i. [Online]. Available: www.woodheadpublishing.com
10. C. Istook, Three-Dimensional Body Scanning to Improve Fit, in *Advances in Apparel Production* (2008), pp. 94–116. https://doi.org/10.1533/9781845694463.1.94
11. S. Hayes, J. Mcloughlin, Technological Advances in Sewing Garments (2008). [Online]. Available: https://api.semanticscholar.org/CorpusID:113837558
12. E. Alho, Computer-aided colour matching of apparel fabric, *Advances in Apparel Production* (Feb. 2008), pp. 26–57. https://doi.org/10.1533/9781845694463.1.26
13. W.R. Kennon, *Developments in Pressing Technology for Garment Finishing* (2008), pp. 250–265. https://doi.org/10.1533/9781845694463.2.250
14. C. Ghituleasa, Supporting digitalization in garment engineering through virtual prototyping, in *Advanced Textiles for a Better World* (2019). https://www.researchgate.net/profile/Olaru-Sabina/
15. P.R.M. Jones, G.M. West, D.H. Harris, J.B. Read, The loughborough anthropometric shadow scanner (LASS). Endeavour **13**(4), 162–168 (1989). https://doi.org/10.1016/S0160-9327(89)80004-3
16. K. Liu et al., 3D interactive garment pattern-making technology. CAD Comput. Aided Des. **104**, 113–124 (2018). https://doi.org/10.1016/j.cad.2018.07.003
17. H. Obaidullah, Impact of computer and its modernization in fashion world. Int. Res. J. Modernization Eng. Technol. Sci. **03**(07), 5–14 (2021)
18. Lectra, Gerber AccuMark & Gerber AccuNest. [Online]. Available: https://www.lectra.com/en/products/gerber-accumark-accunest-fashion. Accessed 24 Feb 2024
19. Gerber AccuMark CAD software for fashion design | Lectra. [Online]. Available: https://www.lectra.com/en/products/gerber-accumark-accunest-fashion. Accessed 25 Feb 2024
20. J.Y. Sun, L. Cen, L.N. Zhai, G. Bai, Research on the design of protective clothing for industrial robots assisted by CLO 3D. Des. Stud. Intell. Eng. **347**, 62–69 (2022). https://doi.org/10.3233/FAIA220011
21. C.V. Fashion, CLO 3D. [Online]. Available: https://www.clo3d.com/en/clo. Accessed 24 Feb 2024
22. CLO Virtual Fashion Ships Marvelous Designer 9.5 | CG Channel. [Online]. Available: https://www.cgchannel.com/2020/04/clo-virtual-fashion-ships-marvelous-designer-9-5/. Accessed 25 Feb 2024
23. P. Jana, Automation in sewing technology, in *Automation in Garment Manufacturing* (Elsevier Ltd, 2017), pp. 199–236. https://doi.org/10.1016/B978-0-08-101211-6.00009-4
24. AP-876|Automatic Machine| JUKI Industrial Sewing Machine. [Online]. Available: https://www.juki.co.jp/industrial_e/products_e/apparel_e/automatic_e/detail.php?cd=AP-876_E. Accessed 25 Feb 2024

25. A. Noris, T.H. Nobile, N. Kalbaska, L. Cantoni, Digital fashion: a systematic literature review. A perspective on marketing and communication. J. Glob. Fash. Market. **12**(1), 32–46 (2021). https://doi.org/10.1080/20932685.2020.1835522
26. M. Fromhold-Eisebith, P. Marschall, R. Peters, P. Thomes, Torn between digitized future and context dependent past—how implementing 'Industry 4.0' production technologies could transform the German textile industry. Technol. Forecast. Soc. Change **166**, 120620 (2021). https://doi.org/10.1016/J.TECHFORE.2021.120620
27. S. Ma et al., An artificial neural network chip based on two-dimensional semiconductor. Sci. Bull. (Beijing) **67**(3), 270–277 (2022). https://doi.org/10.1016/j.scib.2021.10.005
28. İZ. Akyurt, Y. Kuvvetli, M. Deveci, Enterprise resource planning in the age of Industry 4.0: a general overview. Logistics **4**, 178–185 (2020). https://doi.org/10.1201/9780429327636-21
29. G. San-Payo, J.C. Ferreira, P. Santos, A.L. Martins, Machine learning for quality control system. J. Ambient. Intell. Humaniz. Comput. **11**(11), 4491–4500 (2020). https://doi.org/10.1007/S12652-019-01640-4/METRICS
30. W.K. Jung et al., Appropriate smart factory for SMEs: concept, application and perspective. Int. J. Precis. Eng. Manuf. **22**(1), 201–215 (2021). https://doi.org/10.1007/S12541-020-00445-2/TABLES/5
31. K. Liu et al., An evaluation of garment fit to improve customer body fit of fashion design clothing. Int. J. Adv. Manuf. Technol. **120**(3–4), 2685–2699 (2022). https://doi.org/10.1007/S00170-022-08965-Z/METRICS
32. S. Ahmad, S. Miskon, R. Alabdan, I. Tlili, Towards sustainable textile and apparel industry: exploring the role of business intelligence systems in the era of Industry 4.0. Sustainability 2020 **12**(7), 2632 (Mar. 2020). https://doi.org/10.3390/SU12072632
33. R. Nayak, T. Panwar, L. Van Thang Nguyen, Sustainability in fashion and textiles: a survey from developing country. Sustain. Technol. Fashion Text. 3–30 (Jan. 2020). https://doi.org/10.1016/B978-0-08-102867-4.00001-3
34. L. Keßler, S.A. Matlin, K. Kümmerer, The contribution of material circularity to sustainability—recycling and reuse of textiles. Curr. Opin. Green Sustain. Chem. **32**, 100535 (2021). https://doi.org/10.1016/J.COGSC.2021.100535
35. R.J. Geldermans, Design for change and circularity—accommodating circular material & product flows in construction. Energy Procedia **96**, 301–311 (2016). https://doi.org/10.1016/J.EGYPRO.2016.09.153
36. M. Koszewska, Circular economy—challenges for the textile and clothing industry. Autex Res. J. **18**(4), 337–347 (2018). https://doi.org/10.1515/AUT-2018-0023
37. T. Rayna, L. Striukova, Involving consumers: the role of digital technologies in promoting 'prosumption' and user innovation. J. Knowl. Econ. **12**(1), 218–237 (2021). https://doi.org/10.1007/S13132-016-0390-8/TABLES/4
38. P. Brown, B. Baldassarre, J. Konietzko, N. Bocken, R. Balkenende, A tool for collaborative circular proposition design, J. Clean Prod. **297** (May 2021). https://doi.org/10.1016/J.JCLEPRO.2021.126354
39. C. Wijethilake, B. Upadhaya, T. Lama, The role of organisational culture in organisational change towards sustainability: evidence from the garment manufacturing industry. Prod. Plan. Control **34**(3), 275–294 (2023). https://doi.org/10.1080/09537287.2021.1913524
40. G.E.H. Wijewardhana, S.K. Weerabahu, J.L.D. Nanayakkara, P. Samaranayake, New product development process in apparel industry using Industry 4.0 technologies. Int. J. Product. Perform. Manage. **70**(8), 2352–2373 (2021). https://doi.org/10.1108/IJPPM-02-2020-0058/FULL/XML
41. K. Scott, *Fashion at the Intersection of Anthropology, Ecology and Innovation.* https://doi.org/10.1111/ejed.12238
42. L. Suarez-Visbal, C. Stuckrath, J.R. Carreón, Circular economy in the textile and apparel sector: an overview of global trends, challenges and opportunities, Accelerating Sustain. Fashion Clothing Text. 257–267, (Sep. 2023). https://doi.org/10.4324/9781003272878-26/CIRCULAR-ECONOMY-TEXTILE-APPAREL-SECTOR-OVERVIEW-GLOBAL-TRENDS-CHALLENGES-OPPORTUNITIES-LIS-SUAREZ-VISBAL-CLAUDIA-STUCKRATH-JES

43. A New Textiles Economy: Redesigning Fashion's Future. [Online]. Available: https://www.ellenmacarthurfoundation.org/a-new-textiles-economy. Accessed 26 Feb 2024
44. N. Nattrass, J. Seekings, Trajectories of development and the global clothing industry. Compet. Change **22**(3), 274–292 (2018). https://doi.org/10.1177/1024529418768608
45. A. Gangoda, S. Krasley, K. Cobb, AI digitalisation and automation of the apparel industry and human workforce skills. Int. J. Fashion Des. Technol. Educ. **16**(3), 319–329 (2023). https://doi.org/10.1080/17543266.2023.2209589
46. The State of Skills in the Apparel Industry
47. L.A.P. Nathan, S.S. Omar, U.K. Rashid, N. Bujang, Training and developing employees in the Industrial Revolution of 4.0 Era Parit Raja Batu Pahat, Johor, MALAYSIA *Corresponding Author Designation. Res. Manage. Technol. Bus. **1**(1), 899–907 (2020). [Online]. Available: http://penerbit.uthm.edu.my/periodicals/index.php/rmbt
48. Anon, Importance of Training and Development for Employees, Apparel Resources. [Online]. Available: https://online.maryville.edu/blog/importance-of-training-and-development/. Accessed 15 Dec 2021
49. Anon, The Importance of Training Employees: 11 Benefits | Indeed.com, Indeed Editorial Team. [Online]. Available: https://www.indeed.com/career-advice/career-development/importance-of-training. Accessed 12 Jan 2022
50. E. Textile, I. Development, Assessment Practices and Challenges of Training and Development at the Assessment Practices and Challenges of Training and Development at the Ethiopian Textile Industry Development Institute (ETIDI) (March, 2021). https://doi.org/10.7176/JAAS/68-02
51. Anonymously, Interpersonal Skills | SkillsYouNeed. [Online]. Available: https://www.skillsyouneed.com/interpersonal-skills.html Accessed 16 Mar 2022
52. A. Schoar, B. Hyman, M. Desai, *Supervisory Skills Training in the Cambodian Garment Industry: A Randomized Impact Evaluation* (2009)
53. J. Conlon, C. Gallery, Developing digital skills: a fashion business masterclass in virtual 3D prototyping with Style3D, Int. J. Fashion Des. Technol. Educ. (2023). https://doi.org/10.1080/17543266.2023.2247425
54. J.K. Kassah, E. Bruce-Amartey Jnr, R. Acquaye, E. Bruce-Amartey, R.A. Jnr, Fashion design training in technical universities and performance of graduates in computer-aided apparel development in Ghana. Am. J. Art Des. **7**(3), 78–82 (2022). https://doi.org/10.11648/j.ajad.20220703.11
55. P. Vashisht, N. Rani, P. Vashisht, Automation and future of garment sector jobs: a case study of India Standard-Nutzungsbedingungen: Working Paper 385 Automation and Future of Garment Sector Jobs: A Case Study of India INDIAN COUNCIL FOR RESEARCH ON INTERNATIONAL ECONOMIC RELATIONS (2019). [Online]. Available: http://hdl.handle.net/10419/203719
56. *Fashion 4.0. Innovating Fashion Industry Through Digital Transformation*
57. CLO 3D Shaping the Future—Prerit Design Academy. [Online]. Available: https://prerit.org/clo-3d-shaping-the-future-of-fashion/#:~:text=Gone%20are%20the%20days%20of,resulting%20in%20significant%20cost%20savings. Accessed 24 Feb 2024
58. The Future of Fashion—CLO 3D—ABC Seams®. [Online]. Available: https://abcseams.com/clo-3d-virtual-prototyping/. Accessed 24 Feb 2024
59. Robots—A Blessing for Garment Manufacturers. [Online]. Available: https://fashinza.com/manufacturing/industry-4-0/how-are-robots-improving-garment-manufacturing/. Accessed 24 Feb 2024
60. B. (Bart L.) MacCarthy, D. Ivanov, *The Digital Supply Chain*
61. Major Technological Trends in the Garment Manufacturing Industry. [Online]. Available: https://fashinza.com/manufacturing/industry-4-0/latest-technology-solutions-garment-manufacturing-industry/. Accessed 24 Feb 2024

62. 10 Tech Trends shaping the future of textile and garment production | Syndell. [Online]. Available: https://syndelltech.com/top-10-tech-trends-for-the-textile-and-garment-manufacturing-industry/#:~:text=3.-,3D%20Printing%20and%20Digital%20Textile%20Printing,are%20made%2C%20and%20even%20personalized. Accessed 24 Feb 2024
63. F. Bárcia de Mattos, V. Esquivel, D. Kucera, S. Tejani, *The State of the Apparel and Footwear Industry 1*

Chapter 13
Digitalization Impact on Sustainability in Garments Industry

Aqsa Imran

Abstract Digitalization has ushered in a transformative wave in the garments industry, profoundly influencing sustainability practices. One of its most significant contributions lies in fostering transparency across the supply chain. Technologies like blockchain enable the tracing of materials from source to finished product, unveiling unsustainable practices such as labor exploitation and environmental harm. Moreover, digital tools like computer-aided design (CAD) and 3D modeling optimize production processes, reducing material waste by allowing designers to visualize garments digitally before physical production. This shift toward on-demand manufacturing further curtails overstock and waste. Additionally, digital platforms facilitate efficient inventory management, enabling real-time tracking of inventory levels and consumer demand, thus minimizing excess inventory and markdowns. Enhanced communication and collaboration among stakeholders foster collective efforts toward sustainability, while data-driven initiatives leverage digital data to identify areas for improvement in resource utilization and environmental impact. Furthermore, digital platforms empower brands to engage manufacturers in sustainable practices, raising awareness and promoting responsible manufacturing. In sum, digitalization is pivotal in driving sustainability in the garments industry, encouraging transparency, efficiency, and innovation toward a more environmentally and socially conscious future.

Keywords Digitalization · Sustainability · Data driven initiatives · Mass customization · Industry 4.0

A. Imran (✉)
Department of Textile Engineering, School of Engineering and Technology, National Textile University Faisalabad, Faisalabad, Pakistan
e-mail: draqsa@ntu.edu.pk

M. Q. Khan et al. (eds.), *Garment Sizing and Pattern Making*, SDGs and Textiles,
https://doi.org/10.1007/978-981-97-7683-2_13

Introduction

The garment industry or apparel industry is concerned with the designing, construction, distribution, and sale of clothing, apparel, and related items. Due to ever-shifting trends and increase in competition in the global market, manufacturers have started to transform the manufacturing processes to better suit and compete with the recent market. Industry is undergoing a revolution driven by technological advancements that is changing the way clothing is designed and produced, essentially improving efficiency, optimizing process flow, and reducing textile waste [1]. According to the European Technology Platform, the following four concepts will define and shape the textile industry in the upcoming years: advanced materials, digitalization, sustainability, and new growth markets. Additionally, tools and technologies of Industry 4.0 (cognitive technology, data Analytics, AR, VR, 3D prototyping software, cloud computing, IoT, etc.) have proven to be beneficial not just in terms of production but also in identification of supply chain issues and optimization of processes [1, 2].

Impact of Digitalization

The fourth industrial revolution integrates cyber-physical systems, Internet of Things, big data, and other emerging IT concepts with industrial processes. The realization of the industry 4.0 concept has caused immense increase and productivity for the countries adopting it. According to an Oxford University report, automation in manufacturing industries will increase the annual revenue by $4.9 trillion by 2030. The market of automation in the textile industry is expected to grow at the compound annual growth rate of 3.75% by 2028. Due to the rise of IT and related concepts, customers are more inclined to employ online resources to meet their fashion retail needs. The Covid-19 pandemic is also a factor for an increase in e-commerce and online shopping. This rising trend has forced the textile industry to adopt a more digitalized approach throughout the supply chain [3].

Sustainability

The conventional method in textile industries is based on a design, sell, and produce model and can lead to problems in sustainability and production of high amounts of textile waste causing a detrimental effect to the environment. The digitalization in the industry aims to control this issue by adopting a more sustainable approach of selling, designing, and then, producing. This method not only ensures customer satisfaction by providing them with an article that fits their requirements but also guarantees less textile waste production. The process can be further optimized by the employment of automation, mobile technology, robotics, simulations, and digitalization [4, 5].

Sustainability, efficiency, agility, transparency, production flexibility, increased customer satisfaction, reduced operational costs, faster delivery rates, improved production, increased productivity, and market competition are the main driving force behind the rise of automation in the textile industry since 2017. Use of robotics, data analytics, CNC machines, RFID tags, cloud computing, and ERP solutions are the main trends contributing to the automation and digitalization in the industry from 2023.

Lack of skilled labor, privacy and security issues, lack of standardization, unavailability of automation compatible infrastructure in SMEs, and huge investment costs are the challenges being faced by the industry in the path toward automation and digitalization [1, 2].

Industry 4.0 and the Textile Perspective

The rise of digital technology has led to the introduction of an industrial revolution termed Industry 4.0 or Industrial Internet of Things. The technologies introduced by the revolution are also being utilized in the textile sector. Additive manufacturing is being utilized for prototyping [6], AI is being employed for predictive analysis and decision-making [7], augmented and virtual realities are being used in marketing and retail for simulations and communication purposes [8], and autonomous robots, Cobotic systems, and cyber-physical systems are being used to perform tasks with safety issues with greater precision and speed [9, 10]. Additionally, big data, block chain, mobile technology, and cloud are being utilized for data analysis, management, supply chain traceability, and communication [11, 12], while RFID technology and sensors and actuators are being used for tracking, monitoring, and improvement of supply chain performance [13, 14].

Digital Communication

The entire process from design to distribution is based on communication. The current trend in communication is digital information transfer. This not only makes the process easier and global but also transparent and efficient. Wireless sensor networks such as Wi-Fi, GSM or low-power near-field communication technologies, RFID technology, and AutoID sensors are the front runners in achieving this objective. IoT or cloud-based IoT with block chain is an effective source of communication across large distances between the customer and the retailer or between departments. The RFID technology in the production, distribution, and retail departments helps in better communication and enhanced management of multiple processes. It can provide real-time order tracking and assists in real-time monitoring and management. Virtual technologies help in an interactive manufacturing and shopping experience. Big data applications help in data management, storage, analysis, and accessibility.

It also helps in predictive maintenance and self-optimizing decisions and operations [8, 11–14].

The Market Perspective

Automation and digitalization in the marketing department of the textile industry is characterized by digital marketing, augmented reality, virtual reality, AI employment, and 3D prototyping. These techniques help in prototyping, building customer relations, maintaining customer relations, outfit creation, fit analysis, custom look creation, etc. Adoption of digitalized techniques helps retailers in creating an interactive environment, better brand awareness, and creative advertising, while simultaneously providing personalized shopping experience and custom production at an affordable price to the customers—ultimately helping them in smart shopping. Companies such as **Infinity hub** provide digital marketing and marketing automations services to various industries in multiple countries around the world.

Digital Marketing

Some of the tactics performed for this purpose include but are not limited to content personalization, predictive modeling, and customer experience mapping, with the latter being the most effective tactic for optimization in marketing automation. This can be achieved by management tools such as CDP or customer database platform, which implements Enterprise Resource Planning (ERP), customer relationship management (CRM), and propensity to manage and automate marketing processes, or by communication tools such as social media platforms integrated with SEO. A combination of this is the development and advertisement of an SEO-backed company website which collects information such as contact details as well as preferences based on customer consent [3, 15, 16].

Augmented Reality

It combines the real world with digital data to help customers enhance their experience and understand what the product has to offer if used by them in the real environment. Essentially, it is the placement of the digital simulation of the product into the real world and assists in fit analysis and outfit evaluation, without having to try anything on. Tommy Hilfiger has introduced an AR clothing try-on experience in Europe which employs 3D body tracking and product simulations to customers for an interactive shopping experience [8].

Virtual Reality

The creation of a digital world with some real-world elements and the production of simulations such as a virtual fashion show or runway by companies such as CLO and Tuka Tech help in creation of effective brand awareness. The marketing campaigns are focused on creative advertising [8].

Artificial Intelligence

The employment of AI helps in sales forecasting, recognition and evaluation of patterns and trends, and effective and quick solutions to advertisement and marketing campaigns. Algorithms related to data mining help in managing and recognizing customer preferences and then provide recommendations based on those preferences [7, 17]. A digital wardrobe by Alta Daily, for example, is an AI-powered digital solution to both marketing and consumer needs. The AI generates personalized outfits based on existing pieces as well as provides recommendations on what to buy, where to buy it from, and cost evaluation and comparison to help create a custom look. Additionally, consultancies such as "Edited" employ an AI-fueled retail intelligence approach to provide predictability of the market and current trends as well as give more visibility to SMEs in apparel.

3D Prototyping

Digital twin (DT) is a virtual prototyping technique to help visualize the final product and provide insight into the most effective method on its design and production. Therefore, decisions on optimum process selection can be made far before investing in physical prototyping [18]. Additionally, CAD software can be used to first design the product, and then 3D printing techniques can be employed to create prototypes which can be finalized, evaluated, and tested across large distances; thus, customer relations can be built and maintained globally [5, 6, 19].

Digitalization in the Design Department

From 3D modeling to artificial intelligence and from 3D printing to virtual reality, digitalization in the design department is contributing to an optimized production process, creation of innovative and sustainable designs with utmost efficiency. This includes digitalization in the pattern making as well as the marker making process

[20]. This digitalization can be divided into two categories: software/applications and hardware/equipment.

Software and Applications

Designing and prototyping software and applications and virtual simulations make the process much more efficient and easier to evaluate, redesign, and replicate. Examples of these are as follows:

- ***Marvelous Designer***:

A digital clothing, design, and simulation software with interconnected workspace for both 2D and 3D designing, it is an industry standard for creating virtual clothing for VFX, design, and architecture industries. The wide plethora of sculpting, positioning, and sewing tools along with a wide range of virtual models make it a great platform to craft virtual clothing.

- ***CLO 3D***:

For fashion design and simulation, CLO 3D and Browzwear are two industry-leading software platforms tailored for digital clothing design and simulation, offering both 2D and 3D workspaces. CLO 3D, unlike Marvelous Designer, was created specifically for the apparel and fashion industry and thus provides various features essential for the garment production process such as precision points as well as the option to export the design in a variety of format among others.

- ***TUKA 3D***:

TUKA 3D is a virtual prototyping software dedicated exclusively to 3D design, prototyping, and the evaluation of textile products for specialized applications. Additionally, Tuka 3D fit can be used to create aesthetically pleasing presentations to showcase the fit of the created design without the need for physical samples.

- ***TUKA CAD***:

A product of TukaTech, it provides a precise pattern making experience due to its advanced functionalities and process engineering. The software can be used to create pattern designs, 3D patterns' making and grading.

- ***VStitcher***:

It offers realistic simulations of the final product by transforming 2D patterns into 3D garments. The options provided by the vast fabric library can be used to generate realistic simulations of the garment and therefore, test its look, fit, and comfort.

- ***OPTITEX***:

It designs, evaluates, and simulates from both 2D and 3D perspectives, offering models for both humans and furniture. The platform provides features for nesting, grading, pattern making, and costing. The advanced features include the calculation of total consumption—thus reducing time and effort and increasing efficiency.

- ***Tailor Nova***:

It offers visualizations and simulations to generate patterns in 3D and assess the appearance of the final garment. The features of spec menu as well as visualization tool contribute to its excellence.

- ***Audaces Fashion Studio***:

It enables the creation, design, and visualization of garments in a 3D workspace, offering easy-to-use, interactive, and highly realistic simulations. It provides application of realistic simulations for texture, color, and prints and reduces the number of prototyping attempts in the design phase.

- ***Pantone Studio***:

It facilitates the creation of color libraries, patterns, stories, and palettes, while also enabling color extraction and replication from the real world to the virtual environment. The created color libraries and palettes can be tested on 3D-rendered materials and designs. Additionally, the cloud service proves to be a great management and communication feature.

- ***Digital Sketching Software***:

These help in the sketching and drawing process of the designing step. The sketches can be accessed again and again, and corrections can be made easily and effectively. Sketchbook and Prêt-à-Template are two such applications.

- ***Fashion Design Flat Sketch***:

This is helpful in creating technical sketches for tech packs and base sketches for design development. It helps create two-dimensional silhouettes of the final garment and, essentially, produces the "blueprint" of the product.

- ***PAD System***:

It is an all-in-one tool for designing products, as well as creating and grading patterns. The software can be used to create from scratch or to edit digital artwork to craft new designs.

- ***Lectra Diamino Fashion***:

Created specifically for marker creation, it helps in fast creation of effective and efficient markers. The software is made specifically for the creation of effective patterns and markers required in the automotive industry—thus providing a competitive advantage in a field where production volumes per model are often high.

- ***Gerber Accumark***:

It assists in creating patterns and markers, along with automating calculations for consumption, utilization, and efficiency. The Accumark family provides a number of great softwares for the design, development, and production process involved in pattern making.

Hardware/Equipment

The hardware and equipment in are characterized using digital equipment such as drawing tablets, IT tools, and techniques such as cloud computing and physical prototyping such as the combined employment of body scanner (for measurements extraction), CAD software (for design creation), and 3D printing (to convert the virtual design into physical design). This process is also known as Data Twin (DT) [18].

Automation in the Cutting Department

Automations in cutting department aim at optimizing material utilization, reducing manpower, and increasing production and process efficiency. This is achieved using robotics and a combination of machine and human labor or cyber-physical systems. An automated cutting room, such as the one proposed by Lectra, works on the principle of plan, execute, and optimize. Lectra has provided automated cutters for easier mass customization/made to order products, Virga for full single-ply cutting and Vector for multi-ply cutting, for the cutting room to enhance efficiency, optimize the process, provide agility and versatility, and increase profitability [2].

Gerber has also launched Paragon cutter HX series integrated with its CAD software Gerber Accumark. Morgan Technica has provided a cutting line with automated cutting and integrated with software for CAD, cut order planning, PDM, and virtual fitting. Last but not the least, Tuka Tech has been able to engineer TukaSpread, capable of spreading, plotting, and cutting in one go and is integrated with 2D and 3D pattern making, design, and manufacturing softwares for designing and automatic plotters, spreader, and cutter for production. A computer-controlled cutting line consists of CNC machine hardware, a table with conveyer belt, a blower system, a computer-controlled knife/cutter, knife sharpening tool, automatic drilling tool, display monitor, sensors, control panel, bundling region as well as spreading region. Additionally, laser-cutting machines are slowly replacing traditional cutting equipment, as it addresses the issue of fraying edges and is a more efficient solution to cutting synthetic fabric. The laser head is controlled by a computer system [21].

Automation in the Sewing Department

The automations in the sewing department are characterized by robotics, cyber-physical systems, and CNC machines. Additionally, the entire system is optimized by the employment of RFID tags for transparent and effective communication from one operator or machine to another and from one department to another [9, 16, 22].

Sewbot by Softwear Automation

With the major objective of achieving sustainability in the textile sector, Softwear automation's Sewbots are a machine vision robotic work line able to sew garments automatically and precisely by performing variable types of stitching, seaming, and cutting operations.

- **Lowry Sewbot**

It employs computer vision and artificial intelligence to adjust and sew garment, like a seamstress.

- **Dürkopp Adler**

It is a semi-automatic machine capable of sewing double welt pockets easily and efficiently.

- **Automatic Surging**

Traditionally, the edges of a garment are sewn separately. The automated surging provides sewing and overlocking the edges to prevent them from fraying in one go.

- **Automated Pocket Attaching Machine**

The machine requires minimal operator intervention and streamlines the entire process of pocket attachment onto garments. The entire process is managed by a single operator.

- **Belt Loop Attachment Machine**

The belt loop attachment is a bottleneck operation. The automated process reduces the number of errors and operators.

- **Sewing Machine with Automated Bobbin Changer**

It works on the automatic analysis of remaining thread in the bobbin and replacement of it with a filled one via the employment of a robotic catcher. It is essentially a CNC machine and enhances productivity.

- **Buttonhole Indexer**

It is a CNC machine equipped with a control panel for inputting data. It enhances the quality and speed of the process.

- **Role of AR**

Augmented reality can be used to train workers without wasting resources and to evaluate operator efficiency. This helps them garner experience in the digital space before they move to the physical environment and thus reduces the chances of operator-based errors [8].

Kinect Technology

Kinet technology helps train cyber-physical systems into mimicking the way a human operator sews a particular garment by first observing and analyzing how an operator sews and then utilizing that data into training the robotic systems [1].

Automation in the Finishing Department

Like automation in the sewing department, the entire process is linked via the employment of RFID tags, that convey information according to customer requirements [9, 16, 22]. Automation in the finishing department of the garments industry involves the integration of advanced technologies to streamline processes and enhance productivity.

Automatic Binding, Button, and Buttonhole Machine

These are semi-automatic machines provided by suppliers such as Juki, Dürkopp, Rimac, and Pegasis. Juki manufactures sewing machines that can sew buttons as well as buttonholes automatically with minimal human intervention. Additionally, Rimac has introduced an automatic machine capable of neatly finishing curved corners of apparel.

Automated Ironing and Packaging

GEMtex lab has combined AI with production processes for improved and more efficient production and finishing processes of ironing and packaging. Additionally, combination of cyber-physical systems and RFID labels are employed for faster packaging.

Quality Control

With the production of computer-aided quality control systems, advanced image processing methods and machine learning approaches streamline the process flow and improve the quality control check on individual garments. They also efficiently report the quality problems, and the RFID tags can then be used to backtrack the process to evaluate the cause of the problem and possible solutions.

Warehouse Management

The integration of cyber-physical systems with RFID tags and wireless network of communication provides real-time warehouse monitoring and management.

Smart Factories

The concept of a smart factory, driven by digitalization, has emerged as a transformative force in the garments industry, significantly impacting sustainability practices. At its core, a smart factory integrates cutting-edge technologies such as the Internet of Things (IoT), artificial intelligence (AI), robotics, and data analytics to create a highly interconnected and automated production environment. In the context of sustainability, smart factories offer several key advantages.

Firstly, smart factories enable real-time monitoring and optimization of energy and resource usage. IoT sensors (RFID) deployed throughout the production process track and material flow, allowing for precise control and efficient utilization of resources. This not only reduces waste and environmental impact but also lowers operational costs.

Secondly, AI and data analytics algorithms analyze vast amounts of data generated by smart factory systems to identify opportunities for sustainability improvements. By analyzing production data, AI algorithms can uncover inefficiencies and suggest optimizations that minimize waste and enhance resource efficiency. This data-driven approach enables continuous improvement and fosters a culture of sustainability within the manufacturing process.

Furthermore, smart factories facilitate agile and responsive production, aligning with the principles of sustainability. By leveraging digital technologies such as 3D printing and flexible manufacturing systems, smart factories can produce smaller batches of garments on-demand, reducing overproduction and minimizing excess inventory. This lean manufacturing approach not only reduces waste but also enables faster response to changing market demands, ultimately enhancing the industry's overall sustainability.

Additionally, smart factories promote transparency and accountability throughout the supply chain. Blockchain technology, for instance, can be used to create immutable records of every step in the garment production process, ensuring transparency and traceability of materials and labor practices. This transparency empowers consumers to make more informed purchasing decisions and encourages companies to uphold ethical and sustainable practices.

In conclusion, the concept of smart factories represents a paradigm shift in the garments industry, leveraging digitalization to drive sustainability. By integrating advanced technologies, optimizing resource usage, fostering agility, and promoting transparency, smart factories are poised to revolutionize the industry's approach to sustainability, creating a more environmentally and socially responsible manufacturing ecosystem.

References

1. E. Gökalp, M. Onuralp Gökalp, P. Erhan Eren, Industry 4.0 revolution in clothing and apparel factories: apparel 4.0. Manage. Inf. Syst. Perspect. **169**. https://doi.org/10.3726/b15120/21.
2. V. Petrovic, M. Pesic, D. Joksimovic, A. Milosavljevic, *Digitization in the Textile Industry-4.0 Industrial Revolution in Clothing Production* (2019)
3. *Optimizing Marketing Automation Survey Summary Report*
4. Institute of Electrical and Electronics Engineers, *SISY 2020 : IEEE 18th International Symposium on Intelligent Systems and Informatics: Proceedings: Subotica, Serbia, September* (2020), pp. 17–19
5. C. Bai, P. Dallasega, G. Orzes, J. Sarkis, Industry 4.0 technologies assessment: a sustainability perspective. Int. J. Prod. Econ. **229**, 107776 (2020). https://doi.org/10.1016/j.ijpe.2020.107776
6. J. Forman, M.D. Dogan, H. Forsythe, H. Ishii, DefeXtiles, in *Proceedings of the 33rd Annual ACM Symposium on User Interface Software and Technology* (ACM, New York, NY, USA, Oct. 2020), pp. 1222–1233. https://doi.org/10.1145/3379337.3415876
7. H. Nouinou et al., Decision-making in the context of Industry 4.0: evidence from the textile and clothing industry. J. Clean. Prod. **391**, 136184 (2023). https://doi.org/10.1016/j.jclepro.2023.136184
8. P. Goel, K. Mahadevan, K.K. Punjani, Augmented and virtual reality in apparel industry: a bibliometric review and future research agenda, in *Foresight*, vol. 25, no. 2 (Emerald Publishing, Apr. 04, 2023), pp. 167–184. https://doi.org/10.1108/FS-10-2021-0202
9. D. Graham-Rowe, Robot tailoring: stitched by the sewbot. *New Sci (1956)* **210**(2817), 46–49 (Jun. 2011). https://doi.org/10.1016/S0262-4079(11)61454-1
10. *How Robots Change the World What Automation Really Means for Jobs and Productivity* (2019)

11. W.S.D. Wan Abdul Ghani, N.Z. Khidzir, T.T. Guan, M. Ismail, Analysis on factors influencing textile cyberpreneur's intention to adopt cloud-based m-retail application. *Procedia Comput. Sci.* **124**, 345–353 (2017). https://doi.org/10.1016/j.procs.2017.12.164
12. M. Hader, D. Tchoffa, A. El Mhamedi, P. Ghodous, A. Dolgui, A. Abouabdellah, Applying integrated Blockchain and Big Data technologies to improve supply chain traceability and information sharing in the textile sector. J. Ind. Inf. Integr. **28**, 100345 (2022). https://doi.org/10.1016/j.jii.2022.100345
13. B. Unhelkar, S. Joshi, M. Sharma, S. Prakash, A.K. Mani, M. Prasad, Enhancing supply chain performance using RFID technology and decision support systems in the industry 4.0–A systematic literature review. *Int. J. Inf. Manage. Data Insights* **2**(2), 100084 (Nov. 2022). https://doi.org/10.1016/j.jjimei.2022.100084
14. A. Lui, C.K.Y. Lo, Measuring the impact of radio frequency identification (RFID) technologies in improving the efficiency of the textile supply chain, in *Fashion Supply Chain Management Using Radio Frequency Identification (Rfid) Technologies* (Elsevier, 2014), pp. 187–202. https://doi.org/10.1533/9780857098115.187
15. U. Cebeci, Fuzzy AHP-based decision support system for selecting ERP systems in textile industry by using balanced scorecard. Expert Syst. Appl. **36**(5), 8900–8909 (2009). https://doi.org/10.1016/j.eswa.2008.11.046
16. *Digitalization in the Apparel Manufacturing Process* (2018)
17. Y. Xu et al., ChatGPT for textile science and materials: a perspective. Mater. Today Commun. **37**, 107101 (2023). https://doi.org/10.1016/j.mtcomm.2023.107101
18. M.D. Alam, G. Kabir, S. Mirmohammadsadeghi, A digital twin framework development for apparel manufacturing industry. Decis. Analytics J. **7**, 100252 (2023). https://doi.org/10.1016/j.dajour.2023.100252
19. J. Pei, The production system of mass customization using 3D body scanning technology, in *Digital Manufacturing Technology for Sustainable Anthropometric Apparel* (Elsevier, 2022), pp. 185–209. https://doi.org/10.1016/B978-0-12-823969-8.00005-8
20. Y. Xu, S. Thomassey, X. Zeng, An application of machine learning to marker prediction in Garment industry, in *Proceedings of the 3rd International Conference on Applications of Intelligent Systems* (ACM, New York, NY, USA, Jan. 2020), pp. 1–5. https://doi.org/10.1145/3378184.3378219
21. D.B. Datta, B. Agrawal, Computer aided cutting in Indian garment industry: a change agent. *J. Text. Eng. Fashion Technol.* **5**(1) (Jan. 2019). https://doi.org/10.15406/jteft.2019.05.00176
22. R. Pal, A. Jayarathne, Digitalization in the textiles and clothing sector, in *The Digital Supply Chain* (Elsevier, 2022), pp. 255–271. https://doi.org/10.1016/B978-0-323-91614-1.00015-0

The manufacturer's authorised representative in the EU is Springer Nature Customer Service Centre GmbH, Europaplatz 3, 69115 Heidelberg, Germany. If you have any concerns regarding our products, please contact ProductSafety@springernature.com

Printed and bound by CPI Group (UK) Ltd, Croydon, CR0 4YY
08/10/2026
02229948-0001